CHINA'S RURAL MARKET DEVELOPMENT IN THE REFORM ERA

Ashgate Economic Geography Series

Series Editors:
Michael Taylor, Peter Nijkamp and Tom Leinbach

Innovative and stimulating, this quality series enlivens the field of economic geography and regional development, providing key volumes for academic use across a variety of disciplines. Exploring a broad range of interrelated topics, the series enhances our understanding of the dynamics of modern economies in developed and developing countries, as well as the dynamics of transition economies. It embraces both cutting edge research monographs and strongly themed edited volumes, thus offering significant added value to the field and to the individual topics addressed.

Other titles in the series:

The Emerging Economic Geography in EU Accession Countries
Edited by Iulia Traistaru, Peter Nijkamp and Laura Resmini
ISBN 0 7546 3318 7

Urban Growth and Innovation
Spatially Bounded Externalities in the Netherlands
Frank G. van Oort
ISBN 0 7546 3867 7

The Influence of the World Bank on National Housing and Urban Policies
The Case of Mexico and Argentina During the 1990s
Cecilia Zanetta
ISBN 0 7546 3491 4

The Future of Europe's Rural Peripheries
Edited by Lois Labrianidis
ISBN 0 7546 4054 X

The Sharing Economy
Solidarity Networks Transforming Globalisation
Lorna Gold
ISBN 0 7546 3345 4

China's Rural Market Development in the Reform Era

HIM CHUNG
Hong Kong Baptist University

LONDON AND NEW YORK

First published 2004 by Ashgate Publishing

Reissued 2018 by Routledge
2 Park Square, Milton Park, Abingdon, Oxon OX14 4RN
711 Third Avenue, New York, NY 10017, USA

Routledge is an imprint of the Taylor & Francis Group, an informa business

First issued in paperback 2018

A Library of Congress record exists under LC control number: 2004002378

ISBN 13: 978-0-815-38804-3 (hbk)
ISBN 13: 978-1-138-61907-4 (pbk)
ISBN 13: 978-1-351-16176-3 (ebk)

Contents

List of Illustrations		*vi*
Preface		*viii*
Acknowledgements		*xiii*
Note on Measures		*xii*
1	Market-place, Geography and Space	1
2	China's Rural Market District: The Central Place Theory and Skinner	21
3	China's Rural Market Development Since 1979: Has the Traditional Market System been Revived?	35
4	Local Market Pattern and Hierarchy: Case Studies of Deqing and Dongguan	60
5	Beyond the Preoccupation of Transport: Economic Factors and Their Impact on China's Rural Market Development	91
6	Invisible Hand versus Invisible Wall: Administrative Parameters and Rural Market Development	121
7	Deqing, Dongguan and Spatial Variations	148
8	Skinner, Rural Market Development and Economic Reform: A Conclusion	161
Appendix		*167*
Bibliography		*170*
Index		*185*

List of Illustrations

Figures

1.1	Location of Deqing and Dongguan in Guangdong Province	14
2.1	Regular central place hierarchy	23
2.2	Skinner's models of different topology	26
3.1	China's administrative hierarchy since 1982	43
3.2	Proportion of rural wholesale markets by type, 1997	55
3.3	A breakdown of turnover value for rural wholesale markets, 1997	55
4.1	Rural markets and settings in Deqing, 1997	73
4.2	Population, number of markets and turnover value in Deqing	75
4.3	Market classification in Deqing and Wangniudun, 1997	82
4.4	Rural market system in Wangniudun town, Dongguan, 1997	86
5.1	Peasant-market interactions under the impact of high self-sufficiency and low incomes in Deqing, 1997	95
5.2	Peasant-market interactions under the impact of economic factors in Wangniudun, 1997	106
5.3	Peasant-market interactions under Deqing's simple transport network, 1997	112
6.1	Peasant-market interactions in Deqing, 1997	123
6.2	Peasant-market interactions under the impact of administrative factors in Wangniudun, 1997	129
6.3	Peasant-market interactions under Deqing's single market structure and government regulation, 1997	135
6.4	Toll stations and their impact on Deqing's peasant-market interactions, 1997	142

Tables

1.1	Deqing and Dongguan, 2000	15
3.1	The number of rural markets established since the reform era	46
3.2	Rural market turnover value, 1979-2000	47
3.3	Major wholesale markets for agricultural products in China	54
4.1	Arable land and paddy yield in Deqing county	66
4.2	Number of markets created during different time periods in Deqing	71
4.3	Market days at different places in Dongguan city	77
4.4	Market density and village-to-market ratios for Deqing and Dongguan	79
4.5	Village-to-market ratios of different towns in Deqing county	80
5.1	Mr Lu's tax assignment (in grain) in 1997	93
5.2	Mr Ou's tax assignment (in grain) in 1995 and 1997	93
6.1	Toll rate in Wangniudun, Dongguan, 1997	141
6.2	Toll rate in Deqing county, 1997	141

Plates

3.1	The main entrance of a rural market	52
3.2	A typical rural market	52
3.3	Street markets still exist in some remote areas	53
3.4	A scene at a rural market	53
4.1	Decheng central market during a market day and a non-market day	76
4.2	Decheng central market (an intermediate market in Deqing)	83
4.3	A town-level market (standard market) in Deqing	83
4.4	A village-level market (standard market) in Wangniudun	87
4.5	Wangniudun market situated in the town-government seat	87

Preface

My concern with the social implications of local economic geography has drawn me to G. William Skinner's work. Being trained as an anthropologist, Skinner has demonstrated a great interest in traditional Chinese society and its changes. His studies on China's rural markets and urban systems raised heated discussions among geographers, anthropologists and sociologists during the 1960s and 1970s and exerted a tremendous influence on later research. Employing a functional perspective, Skinner pointed out the significance of the rural marketing system in shaping social and economic structures. He stressed that China's rural market is not only a basic unit for commercial functions, but also a central place for social and cultural activities. His analysis of different levels of marketing places – standard, intermediate, central market towns – demonstrated how rural and urban areas were organised in functional terms. This challenged the conventional perspective which emphasises administrative hierarchy and its functions in organising Chinese society. Skinner's studies provided a prototype of traditional Chinese society. While city and countryside in China have experienced dramatic changes under the Communist regime and a planned economy, Skinner's studies show, in a systematic and innovative way, the underlying and unyielding requirements for rational marketing districts and networks that could not be constrained by administrative fiat. This explains why Skinner's studies are frequently mentioned and discussed.

During the 1970s, Skinner further elaborated his thesis of rural marketing districts into a 'regional-systems' theory. He argues that standard-market towns/systems were the basis of an economic hierarchy during Imperial China. These local systems were tied to higher-level systems. From bottom to top, there were a number of levels of central places/systems – intermediate, central market towns, local and regional cities, regional and central metropolises. These economic hierarchies formed eight macro-regions in Imperial China. The eight macro-regions were relatively independent and different from each other because their interregional activities were constrained by physical barriers and high transport costs. Accordingly, Skinner suggested that the macro-region should be used as the basis for historical studies. This perspective extended historians' analytical horizons beyond the traditional practice of using administrative units, such as the province and county, for understanding Chinese history. It also provided a clear distinction between China's administrative and economic hierarchies.

Skinner's mid-1960s research has had considerable influence over the past decades because he introduced a systematic and comprehensive framework to explain Chinese society and its development. In particular, his employment of 'central place theory' and 'historical cycles' to analyse China's rural market system and social changes challenged the 'theory-less' tradition of Sinology. This contribution has established Skinner as a 'master' in China studies.

In this book I have been concerned with the rural market – the basis of Skinner's regional systems. Skinner's series of papers on the topic – "Marketing and Social Structure in Rural China", Parts I, II and III – are regarded as a classic in China studies. After these papers were published in the mid-1960s, his models were widely discussed and were applied to Latin America, Africa and other developing areas (see Smith, 1976a, 1976b). However, applying this framework to study Chinese sites was delayed because of the Cold War during the 1960s and 1970s and political instability in China. This forced Skinner's followers in China studies to illustrate his models with reference to Taiwan, the largest Chinese community outside mainland China (Knapp, 1971; Crissman, 1976a, 1976b; Sangren, 1980, 1985). Given Taiwan's historical and cultural context, Skinner's models hardly found any support there. For example, Crissman (1976a) discovered that the periodic market, as defined by Skinner, had never existed in the island. Sangren's (1980) study in central Taiwan further demonstrated that market growth did not necessarily follow the evolutionary process suggested by Skinner.

The reforms in China since 1978 have provided the first opportunity for scholars to examine Skinner's original framework in the People's Republic of China. For example, Chiu and Leung's (1983) study on periodic markets during the last days of the commune systems provides a testimony to Skinner's work, although no direct comparison has been made. Also, Hodder (1993) examines China's private trading network and attempts to extend Skinner's framework beyond rural market systems. Recently, Skinner's mid-1960s work has been further analysed by Lin (1997a) and Wang, Mingming (1997). Lin uses Skinner's hierarchical framework to understand the function of China's small towns and their development. Wang, in contrast, puts Skinner into an historical context to explore the development of social anthropology in China. In 1998, Skinner's mid-1960s classic was translated into Chinese for the first time by the Chinese Social Science Press (*Zhongguo Shehui Kexue Chubanshe*). This translation has spread Skinner's thesis among contemporary China researchers.

Wang, Mingming's (1997) review of Skinner's contribution in China studies has given rise to discussions on the applicability of Western economic and social theories in China. This is because China was never a part of Western culture and its capitalist economy. This issue does not occur only in Skinner's work, but also in that of scholars who believe that China's social and economic processes can be explained by theories generated through Western experiences. The historical and cultural variations between the East and West have encouraged other contemporary scholars to rethink the validity of Western theories on non-Western societies in general and Skinner's framework in particular.

My work provides a detailed re-examination of Skinner's thesis of rural market development during the era of economic reform. I use Skinner's (1964, 1965a, 1965b, 1977c, 1985a, 1985b, 1994) studies on China's rural marketing during different time periods as the basis of my analysis. Unlike Skinner, whose studies were based heavily on documentary evidence, on-site fieldwork at two places was used in my research. The aim is to identify patterns that vary from

Skinner's prototypes by providing detailed analyses of China's rural market development in different places. Also, this study contrasts current market developments and their spatial distribution, patterns, hierarchy, and underlying causes with Skinner's mid-1960s models and hypotheses. It provides an elaboration and rethink of Skinner's original thesis.

This re-examination of Skinner's thesis takes place within the context of economic reform in China. Particular attention is given to China's changing political economy and its impact on local marketing activities. My findings aim to adjust the pure economic analysis adopted by Skinner and by economists. My emphasis on both political and administrative factors reflects the situation of China, which is politically dominated by one party and the persistence of a highly centralised administrative hierarchy. A distinguishing feature of this study is the illumination of variations between events in the real world – China – and an elegant theory. It aims to demonstrate the inadequacy of grand theories and to rethink their legitimacy in social science studies.

Methodology

This study is based on data collected over 10 months of intensive fieldwork in Deqing and Dongguan, south China, supported by library research in Hong Kong. Unlike Skinner, whose research depended heavily on documents, my fieldwork provides an in-depth understanding of China's rural marketing activities and their operational structure. First-hand data were collected through observations, interviews and household surveys. Dozens of photographs were taken and field notes written to document observations. I visited all 15 towns/townships in Deqing county and most villages in the Wangniudun town of Dongguan. I also visited all rural markets in both areas. In Deqing, regular visits were made to Guanxu, Huilong and Jiushi rural markets to observe the activities of peasants and trading situations during market and non-market days, and in both the farming and slack seasons.

Between 1998 and 1999, household surveys were conducted in Deqing and Wangniudun to examine interactions between peasants and rural markets. The number of interviews was not pre-determined and visits continued until a complete picture of peasant life was exposed. In Deqing, 40 villages in different locations were visited and over 50 households were interviewed. There were three criteria in choosing villages. They had to be: (a) located close together, but under the control of different administrative jurisdictions; (b) served by roads; and (c) situated far away from trunk roads. A similar exercise was also conducted in Wangniudun, with 3 large villages being visited and 10 households being interviewed.

While villages were selected by their location, households were basically picked at random. I was given a free hand in conducting interviews in both field areas, and was assisted by my drivers/interpreters in Deqing and Dongguan – a local road maintenance worker and a high-school teacher respectively. I was well received by the peasants. On occasions, however, my approach was rejected because some peasants were suspicious about talking to strangers. Also, some thought I should interview the village head (*cun zhang*), not them, because he represented the local community. A questionnaire that covered enquiries about the daily lives of peasants and their connections with rural markets was used as a guideline. However, dialogues with peasants often went beyond these pre-set questions. The interviews were conducted voluntarily, though imported cigarettes were given to the respondents afterwards. Generally, people were simple and honest, and were desperate to talk about their lives. Once the conversation started, they readily discussed their life experiences, difficulties and expectations. However, some peasants in Wangniudun were more cautious when talking about their lives. This was particularly the case when sensitive issues were mentioned such as purchase prices and tax burdens. The length of interviews varied from two hours up to a day. Particular households were revisited to follow up on their marketing activities.

Government officials in towns and counties were also interviewed. This was the prime method for obtaining the most up-to-date and detailed information on rural market development. Also, it was the best means of examining the role of local government in rural market growth. Connections were established through banquets organised by the Roadway Bureau, the institution with which I was affiliated. The head of the Bureau introduced me to his colleagues in the Industry and Commerce Management Bureau (ICMB) and other government authorities. Despite connections being established, telephone calls had to be made to verify my identity before interviews took place. No questionnaire was used and dialogues concentrated on three aspects: (a) the history and current situation of the local markets (see Appendix); (b) the way in which the government managed local market places; and (c) the future development of rural markets. Unlike chatting to peasants, interviewing local officials was usually undertaken in a solemn and uncomfortable atmosphere. As anticipated, they spoke about the perspectives of their institutions and vigorously defended state policies. Sensitive questions were sometimes ignored. A follow-up interview was conducted in the summer of 2003 to investigate the latest changes of marketing activities.

A considerable amount of data from secondary sources, such as statistics and documents, were collected to supplement first-hand data. These materials provided the major sources for sketching macro economic development in the chosen study areas. Moreover, they were important for cross-checking information obtained through interviews. Data were mainly obtained from both published and unpublished statistical yearbooks. Published yearbooks were not difficult to find in local bookshops and libraries. With the assistance of the Roadway Bureau, I was able to collect unpublished materials in Deqing, which are regarded as 'internal references' and distributed to people with appropriate authorisation. However, I was warned not to take the materials outside China, not even into the Hong Kong Special Administrative Region (SAR). A similar situation was experienced in Dongguan, but there were no restrictions on carrying the materials abroad.

Supplementary data were also located in local gazettes, newspapers, government documents and even investment brochures.

Library research in Hong Kong allowed me to collect a large number of Chinese books, journal articles and monographs published during the 1990s. These provided valuable information on the different research perspectives of Chinese academics on rural market development.

It is worth mentioning the reliability of statistics used in this book. The legitimacy of Chinese data has always been questioned by researchers. This is because cheating always occurs at different levels of local government. In Deqing, I was able to access different data sources but local officials suggested to me that I inflate some economic figures by 20 per cent for publication. Obviously, such a practice is driven by their desire to show off economic achievements. These problematical figures are not used in this study. Occasionally, a similar set of data was found in other sources, such as newspapers and government reports, but they were equally unreliable because the source was unknown. Moreover, given limited fieldwork resources, it was impossible to conduct large-scale social and economic surveys similar to those undertaken by the Statistical Bureau. Thus, this research uses official figures published in the statistical yearbooks because they are the only complete and comparable data available.

Acknowledgements

This book was made possible by the assistance of many people in Australia, China and Hong Kong. In Australia, sincere gratitude goes to Professor Peter Rimmer for his insightful comments and patience in reading numerous drafts of this manuscript. His all-round support is gratefully acknowledged. I am also grateful to Professor Jonathan Unger for his intellectual stimulation. His positive remarks and encouragement sustained me through many difficult times during the long research and writing process.

In undertaking my fieldwork in China, thanks are due to Steve Yuan and 'Uncle Sum', who helped establish various connections in Deqing and Dongguan; and to all personnel in the Deqing Roadway Bureau and the Wangniudun Town Government for their kind assistance. Moreover, I would like to extend my appreciation to all interviewees from Deqing and Dongguan, who not only gave me their time but also their insights on many rural issues.

In Hong Kong, I would like to acknowledge Betty Kim, Fanny Chu, Karen Ho and other former colleagues at the Chinese University of Hong Kong for their research assistance during various stages of the project; Drs T. Fung and W.-S. Tang for their advice on maps and fieldwork issues respectively; and Mr S.-L. Too of the Geography Department of the Chinese University of Hong Kong, who drew the excellent base maps for this book.

An immeasurable amount of thanks is given to Barbara Banks for copy-editing the manuscript. C.Y. Kwok prepared the book's index and Vivian Kwok assisted on the early stage of this project. Their contributions are gratefully acknowledged.

Portions of the materials in Chapters 4 and 6 have drawn on my pervious published work in *Journal of Transport Geography* (2002). I am grateful to the publisher for permission to make use of these materials.

Last, but by no means least, I am deeply indebted to my wife, Eunice, for her wholehearted support on this project. She played mother as well as father to our baby boy when I was either away on fieldwork or working in the office till late at night – yet her love and inspiration were always with me and motivated me to work hard to complete this project. It is to her, and our son Alvin, that I dedicate this book.

To Eunice and Alvin

Chapter 1

Market-place, Geography and Space

23 May, 1998, Deqing

It was a big day for Mr Li[1]*, who had just seen his maize mature and was ready to sell at the rural market. Living on the edge of Xijiang River, Mr Li cultivated a variety of crops for his own consumption and sale. Taking advantage of the fertile alluvial soil and the dry season, he grew maize in the riverbed which is submerged during the rainy season. Good weather during the past few months had resulted in a high yield this year – about 300 jin – so Mr Li anticipated greater returns at the market.*

As on busy farming days, Mr Li woke up at day break. He cut some maize and put it in the bamboo basket hanging on his bicycle. Around seven o'clock in the morning, he arrived at Jiushi market which is situated 5 km east of his home. There were already dozens of peasants selling similar products in the market-place but Mr Li was not surprised at this. He was confident about his products and set his price at 0.4 yuan per jin, which was quite low, expecting that this would attract customers.

The rural market was not as busy as usual, even though this time of the year was generally regarded as the slack season in Deqing. While waiting for customers, Mr Li chatted with other peddlers. These conversations were probably the only way to exchange market information. Most customers – urban residents living in Jiushi town-government seat and peasants from nearby villages – came after nine o'clock. They liked to negotiate a lower price but not many of them accomplished a satisfactory deal. There were uniformed Industry and Commerce Management Bureau officials patrolling the market-place and levying fees – a combination of administrative costs and rent – from peddlers. Occasionally, peddlers were asked to move if they blocked walkways. Peddlers, mostly peasants, were afraid to offend them because they had jurisdiction over the market-place.

During the afternoon the market was much quieter, despite the rain stopping. However, Mr Li did not leave until four o'clock. After nine hours in the market, only half of his maize was sold. After deducting the market fee, only 20 yuan profit was made. Looking at the unsold maize, Mr Li was a bit worried. After a couple of cigarettes, he decided to take the maize to Decheng Central Market the following day, the county's only daily market which is situated 20 km from his home. The longer distance and the presence of a toll station implied transport costs and toll fee would be higher. However, hoping that the maize would be bought by restaurant buyers and mobile traders, Mr Li said he did not mind paying if he could sell his products.

---

[1] The name of all peasants in this book has been changed to avoid identification.

This study investigates the relationship between the Chinese state and economic forces in shaping rural market development during the reform period since 1979. Economic reform has changed China's economy from a rigid, planned system to a more market-oriented one. Initially, China's economic reform was guided by the paradigm of the 'birdcage economy' which emphasised the planned economy as primary and the market system as secondary (Yabuki, 1999). During the 1980s and 1990s, this model was transformed into the 'market adjusted planned economy' and the 'socialist market economy' that placed more emphasis on market forces and guidance.[2] In rural areas, this shift resulted in peasants having more control over agricultural production and marketing. Consequently, agricultural performance and productivity have increased and the number of rural markets and their turnover value have grown markedly.

Neither economic reforms nor membership of the World Trade Organisation (WTO), however, have given free rein to market activities in the countryside, despite the rapid growth in the number of outlets. Rural markets are established and managed by government. Important crops, such as rice, cotton and other special local products, are still subject to state procurement and price control. Although marketing channels have been opened, peasants still find it difficult to find outlets for their crops (Chen, 1997; Mei, 1998; Gong, Kang, Su and Wu, 2001). Further, the use of trade barriers to regulate commodity flows during the 1980s, and the reissuing of grain coupons in big cities during the 1990s, prove that government regulation still plays an active role in marketing activities. Moreover, the level of market liberalisation varies between places (Gong, 1997; Ding, 1998; Xie, 1998).

The considerable scholarly attention given to China's rural market development predates the Pacific War.[3] In 1940, Spencer published probably the earliest paper in the western world on China's periodic markets. Focusing on Sichuan province, Spencer (1940) provided a detailed description of the area's rural market fair, its operations and schedules. Then, Yang (1944) produced another detailed study on Shandong Province's periodic markets, and their relations with the regional economy, based on fieldwork conducted in the 1930s. The most marked advance followed Skinner's (1964, 1965a, 1965b) application of central place theory to the rural market system in the Chengdu plain of Sichuan

---

[2] See "The State to control the market, the market to lead enterprises" – Zhao, Ziyang's report to the 13th Party Congress, 1987 and "Decision on some issues concerning the establishment of a socialist market economy", 1993.

[3] Rural market refers to market-place – the physical locus for goods and services exchange – in most of this book except when otherwise indicated.

province. Particular attention was paid to the hierarchy of market-places, their distribution, evolution, and relationship to the administrative system. Skinner's perspective on how rural China is organised and how the market system is supposed to change over time has become the classic work in geographical studies on China.

Chinese historians have also demonstrated their great interest in rural market development. Based on over 400 local gazettes and historical materials, Gao (1985) and Li (1989) have studied this system in Sichuan and Guangdong respectively. Gao carefully examined Sichuan's periodic markets and their characteristics – market days, pattern, hierarchy and functions – during the mid-Qing (1700-1800) dynasty. He argued that periodic market systems, rather than rural households, were the basic components of Sichuan's regional economy. Li's (1989) study on Guangdong's markets between the mid-Ming (1500s) and mid-Qing dynasties highlights rural market development during different imperial empires. His detailed examination of changes in market numbers, taxes levied and local government management also shows the rise and fall of periodic markets in Imperial China. Other historical studies on this subject include: Wu's (1983) work of national market hierarchy during the Ming dynasty (1368-1644); Cong's (1995) examination of specialised market towns between the late 1800s and 1930s; and Zhu's (1997) general analysis of rural markets and their relationship to modern transport technology during the early twentieth century, with a particular emphasis on railways.

During the reform period, rural markets as an area of conflict between state control and market forces have also attracted the attention of scholars. Many of them have been concerned with the reforms and their impact on marketing activities in the countryside. Economists, notably Findlay, Watson and Martin (1993), have examined policy shifts since 1978 and their impact upon the country's rural market activities. By analysing the state distribution system since 1978, Watson (1988) provides a detailed study of market changes during the reform era. Based on the distinct growth in rural market numbers and their turnover value, these studies have regarded the liberalisation of rural markets in 1983, the abolition of unified purchases and sales in 1985, and the deregulation of the price system in early 1990s, as a complete victory for market forces. Consequently, they argue that the traditional market system, driven by market forces, has been revived. Skinner (1985a, 1985b) has also joined the post-reform discussions by making a similar contention in his mid-1980s papers, work which completed his study of rural market development during different periods of China's history.

Despite this matter having been addressed over time, these earlier works have been unsatisfactory. Although studies by geographers have identified spatial structure and hierarchy in rural markets, they have been preoccupied with Imperial China. Chiu and Leung (1983), Liu (1991) and Shi's (1995) papers offer rare contributions on the reform period. In Guangdong, Chiu and Leung (1983) examine the marketing attendance of peasants in Shaping and Gulao communes and document the spatial organisation of periodic markets on the eve of economic reform. Although they observe the significant impact of the commune boundary on people's choice of market, they fail to put it in the Chinese context. Thus, the underlying reasons for the observable morphology have not been fully exposed. Also, the relations between the market pattern and its local socio-economic conditions are not clear. Using the Liucheng city of Shandong and the Baoji city of Shaanxi as case studies, Liu (1991) and Shi's (1995) respective studies produce detailed examinations on market days, hierarchies and spatial patterns. Despite their geographical perspective, their discussions focus on what Smith (1979, p.473) has termed "spatio-temporal synchronisation". This notion assumes there is a positive correlation between market days and market location – the closer their site, the closer their schedules. Although the concept of spatio-temporal synchronisation deals with space, it focuses on the correlation between periodicity and market venue. The external relationships between rural markets and their socio-economic contexts are ignored. Also, variations generated by different rural market levels, size and functions have been oversimplified. Market densities have been overlooked. Moreover, the role of participants – traders, producers and consumers – and their interactions with rural markets have rarely been addressed.

In contrast, the majority of post-1979 rural market studies employed an economic perspective, concentrating on the effects of policy changes on growth and examining the income and consumption power of peasants. Macro economists, notably Watson (1988) and Findlay, Watson and Martin (1993) have highlighted the connections between state policy directions and rural market development. However, they have omitted the impact of policy ignorance and distortion at the local level, and the role of local governments is inadequately acknowledged. Thus, their assessment of rural market development is flawed. Similarly, focusing on policy formulation, Chinese economists employ a normative perspective to investigate the function of rural markets. For instance, Shu (1999) regards them as a key element to achieve long-term domestic growth. A series of measures is suggested to improve the current market system so as to release the huge economic capacity of the Chinese countryside. Du (2001) further argues that China's rural market network and the country's urbanisation process are interdependent. Hence, improvement of the rural market network accelerates the process of rural transformation and, eventually, eliminates the inequality between urban and rural areas. Despite their insight on China's rural market system, their analyses remain superficial. The suggested measures are vague with no serious consideration on the country's structural problems.

Most micro-economic perspectives have been contributed by Chinese scholars, notably Zheng (1999), Jiang and Liu (2000) and Yan (2002). They have argued that the income of peasants and their purchasing power have been the key factors stimulating rural market development; accordingly, future market planning should focus on increasing such income. Still, they failed to recognise that income levels were not the only factors generating market development – transport costs and level of agricultural productivity were also important.

G. William Skinner's mid-1960s economic perspective has provided a more comprehensive framework for studying China's rural markets. However, due to the Cold War between the western and communist countries and the political turmoil within China during the 1960s and 1970s, Skinner's notion on market hierarchy was tested in Taiwan and different results were found (Knapp, 1971; Crissman, 1976a, b; Sangren, 1980, 1985). Economic reform since 1979 has given scholars the first opportunity to justify Skinner's theories in China. Jiang (1993) revisits Skinner's fieldwork site in Sichuan and attempts to identify the changes in the market system since 1949. Although his relatively sophisticated qualitative analysis pinpoints the mathematical mistakes in Skinner's work,[4] the investigation is limited to the spatial pattern and hexagonal networks of rural markets. Hodder (1993) has questioned the significance of such a framework to understand China's marketing activities. He argues "if patterns are looked for, they can be found" (Hodder, 1993, p.36). Therefore, whether a marketing area is circular, square or a hexagonal shape is not important. It is the arbitrary assumption of a causal relationship between patterns and process that is subject to criticism. Investigating private trade and markets during the reform era, Hodder goes on to question the assumptions given to the relationship between market hierarchy and settlement, and the arbitrary division of a rural market system. His critique of Skinner's thesis focus on the latter's perspective – marketplaces are merely an allocation network – and argue that markets are "an expression of the institutionalisation of trade" (Hodder, 1993, p.32). These criticisms not only suggested an alternative to Skinner's dominant framework, but also demonstrated that these studies on China's marketplaces are "dated and limited" (Hodder, 1993, p.31). Indeed, new challenges to the validity of Skinner's thesis have resulted from changes in social, political and economic sectors since reform, which have modified the relationship between rural market development and its local context.

One of the most distinct characteristics of China's reform era has been the decentralisation of economic decision-making power. Indeed, this process has transferred the state's command structure to local government (Chung and Tang, 1997). The decentralisation of power has increased the role of local government which, in turn, has had important repercussions for China's political economy and regional development. Initially, greater autonomy enabled local government to play

[4] Szymanski and Agnew (1981, p.39) have criticised Skinner's application of central place theory as "diagrammatic and without mathematical basis".

a more active role in economic development. Economic strategies were formulated and a variety of models devised such as the Sunan and Wenzhou models. Local government involvement in economic growth has increased diversification between places, while at the same time decentralisation has created local protectionism (Lee, 1998). Consequently, trade barriers and 'commodity wars' between provinces have been witnessed during the 1980s and early 1990s. Protectionism has challenged the paramountcy of the markets, suggesting that a complete dismantling of state control has not occurred in post-reform China.

What has been the impact of the above political and economic changes on rural markets, their patterns, and development during the reform era? These issues are examined here. Particular attention is given to the dynamics of China's political economy and the effects on market formation and patterns. Unlike previous studies of the policy implications of market development, this work concentrates on peasant-market interactions. Peasants are the major participants in rural markets, thus, their behaviour and preferences have been decisive in shaping rural market activities and their growth. The other objective of this analysis includes examining market development at locations with different economic and social structures.

Space and context in the study of China

My concern with China's changing political economy and its impact upon market development underlines the importance of adopting a contextual approach. This method criticises the theoretical oversimplification of most studies and suggests that an analysis of economic activities should not be divorced from their political and social factors (Cox, 1995; Sunley, 1996). Indeed, these factors are "embedded or incorporated into surrounding social and institutional conditions" (Sunley, 1996, p.338). Accordingly, abstract theories cannot explain the specific social, economic and political characteristics of places because they are "unique combinations of attributes" (Cox, 1995, p.308). Any understanding of the geography of China's rural marketing system must recognise specific local contexts, particularly the characteristics of the planned economy and the changes that have occurred during the reform era.

The contextual approach not only reminds geographers to consider other causal mechanisms, but also highlights the importance of space. Recent studies on rural markets have addressed the uneven development between inland and coastal regions, and between rural and urban areas. Drawing different examples from Chongqing, Beijing and other places, Findlay, Watson and Martin (1993), for instance, have discovered that wholesale markets play dissimilar roles in urban and rural areas. Huang and Rozelle's (1998) study of rural consumption markets highlighted their irregular growth across China. Rarely have spatial differentials been examined. Space has always been seen passively as the outcome of other major causal mechanisms, notably transport costs and consumption levels. Generally the explanatory power of space has been ignored.

This argument for paying more attention to space should not lead to spatial determinism. Despite dissatisfaction with the spatial amnesia of most economic and sociology theories and the acceptance of the notion that space makes differences, most studies have only mentioned it in passing. Tang's (1997) review of the literature on China's urbanisation illustrates this point. According to Tang, current studies have acknowledged differences between inland and coastal areas, urban and rural areas, and within cities. However, the 'anti-urbanism', 'class-struggle' and 'economic imperative' perspectives used have failed to understand the importance of variations in spatial differentials, and how they relate to a particular context. As noted by Tang (1997, p.32), most studies have "concentrated on the causal mechanisms of the economy but not on the spatial contingent effect". Given this dissatisfaction, it is imperative to theorise space and its effects on social and economic development in a more sophisticated way, while avoiding spatial determinism.

Current discussions of space have provided significant insights. Generally, the concept of 'absolute space' – an independent entity with its own effects – has been rejected (Urry, 1985; Sayer, 1985, 1992; Sheppard, 1995). In contrast, according to Duncan (1989a, p.132), space:

> only exist[s] as a relation between objects (such as planets, cities or people) which have substance. Space itself does not have substance – there is nothing there – and without objects which do have substance there is no spatial relation.

Causal powers are thus "embodied in societal mechanisms, and it is those mechanisms that give life to space" (Sheppard, 1995, p.291).

Massey (1985, p.11) asserts that "there is no such thing as purely spatial processes; there are only particular social processes operating over space". When geographers discuss the 'friction of distance', they are, in fact, referring to land rents or transport costs and their effects on spatial zones. As a result, land rents and transport costs are functions between landowner and labour, and truck owner and consumer respectively. Obviously, it is social objects which interact, not the space or distance. Thus spatial relations are, in fact, relationships between objects (Urry, 1985; Sayer, 1985, 1992; Duncan, 1989a, 1989b). Moreover, Sayer (1985, p.52) suggests that space makes a difference:

> in terms of the particular causal powers and liabilities constituting it. Conversely, what kind of effects are produced by causal mechanisms depends *inter alia* on the form of the conditions in which they are situated.

It is the difference that space makes which defines the particular context of places. Sayer (1985) believes that this is the contingent effect of space. Unlike causal effects, which bring changes to other objects, contingent effects do not have any causal power. However, they make up the *contextual conditions* which affect the operation of causal powers.

Space, according to Duncan (1989a, 1989b), makes a difference in three ways. The first is the *spatial contingency effect*. Given that space is not an object and has no causal power, it will not cause things to happen but it will "influence how, and to what degree, primary causal objects interact, and hence how processes work" (Duncan, 1989a, pp.138-139). While space itself has no causal power, causal mechanisms do operate under the influence of spatial contingency. This implies such mechanisms are not universal. They are 'spatially-bounded' and occur in a particular time and space. Duncan (1989a, 1989b) has categorised this as the *spatial boundary effect* or *local causal process*, a second level of spatial variation. The difference between spatial contingent effects and spatial boundary effects, as highlighted by Tang (1997, p.26), is that the former "affects *how* changes occur" while the latter "*causes* changes to happen". Another difference is that spatial contingent effects usually refer to universal mechanisms and spatial boundary effects pointing to locally specific causal mechanisms.

Finally, there is a third level of spatial variation: *locality effects*. As stated by Duncan (1989b, p.247), these are:

> [t]he contextual effects of local causal powers and spatial contingency may be so significant as qualitatively to alter the nature of social structures in a particular place and hence social action.

Although locality effects have been identified, they are rarely significant. This is because a "spatially specific system of causality can[not] be implied from the spatial boundary effect" (Tang, 1997, p.26). The significance of locality effects, according to Duncan (1989b), has to be demonstrated empirically, and he warns that they may not exist even though a significant spatial variation is observed. Consequently, such effects can almost be ignored (Tang, 1997, p.26).

The above three concepts provide a useful framework for understanding China's social economic development and the country's highly diversified local situation. At a higher level of abstraction, 'socialism with Chinese characteristics' reminds scholars that China is different from other socialist countries and has its very own context. Although China adopted the Soviet development model in the pre-reform period, it was not a pure duplication. After the first Five Year Plan (1953-1957), China began to develop its own distinctive road toward socialism (Selden, 1989; Liao, Gao and Zhou, 1991; Gregor, 1995; Nathan, 1997). China's system under Mao, as suggested by Nathan (1997, p.54), was:

> [a] hybrid between the original half-realised Stalinist pattern and various other patterns, some drawn from the [Chinese Communist] Party's Yan'an experience, some from the KMT [Kuomintang] and Japanese tradition, some developed as experimental, on-the-spot adjustments to immediate economic or political problems.

In the agricultural sector, for instance, the redistribution of farmlands, aimed at eliminating exploitation of peasants by landlords, was an important measure based solely on Chinese experience (Selden, 1989). The Chinese model also had a more decentralised planning and administrative system than its Soviet counterpart. Under the principle of 'unified planning, subordinated management (*tongyi jihua, fenji guanli*)', a hierarchical administrative system was established and authorities at different levels were endowed with certain degrees of power over decision making, material allocation and personnel assignment. The number of enterprises which operated directly under central control and administration was fewer than in the Soviet model. Moreover, the practices of self-reliance, self-sufficiency and high accumulation in the industrial sector were also without a Soviet precedent. These variations were derived from the country's context, in particular its huge and poor population, and international isolation when the Chinese Communist Party (CCP) came to power.

Differences between China and other socialist countries persisted in the reform era. This is underlined by current debates on the transformation of socialist societies.[5] Economic reform in the Soviet Union and other eastern European countries, such as Poland, Czechoslovakia and Bulgaria, has quickly destroyed the old administrative apparatus and systems. Liberalisation and privatisation have been implemented rapidly to construct a capitalist economy within a short period. In contrast to the 'big bang' approach, reform in China is more cautious. Economy and society have been protected by the gradual liberalisation of prices, material supplies and markets. Politically, the administrative apparatus and systems remain unchanged. Although China's reform has always been criticised for its tiny steps, the outcome, when compared with Russia and other eastern European socialist countries, has surprised many observers. It is beyond the scope of this study to address why China has been more successful in reforming its economy than Russia and other eastern European socialist countries.[6] These variations stem from spatial

---

[5] The debate on the approach of transformation is made between 'big bang' and 'gradualism' theorists. Scholars such as Rawski (1999), Lees (1997) and McMillan and Naughton (1992) generally agree that the gradualism has contributed to the success of China's economic reform. For an overview of the arguments, refer to Murrell (1992).

[6] For a more detailed study on the issue, see Nolan (1995).

discrepancies between China and other socialist countries. Their spatial contingent conditions have created disparate outcomes. Initially, socialism was developed differently in China and the Soviet Union and other eastern European countries. Since the reform era, these socialist, or formerly socialist, countries have adopted a number of approaches to reforming the economy and the results have varied. China's divergent transitory process has formed a special context, or contingent conditions, that triggered off other causal mechanisms.

At another level of abstraction, regional variations between north and south, inland and coastal regions, and between different provinces and counties, have provided spatial boundary effects. Regions have unique economic backgrounds, local histories and social characteristics. For example, rural markets have a number of names because of cultural and language variations within China. In the middle and lower courses of the Yellow River they are known as '*ji*'. However, rural markets are known as '*shi*' and '*xu*' in the Yangzi River Valley and southern China respectively, and in south-western China, they are known as '*chang*' or '*jie*'.

Social and economic processes operating under the influence of contingency effects also created locality effects, including: (a) the role of local government; (b) the local economic structure; (c) the relationship between rural and urban sectors; (d) accessibility; and (e) the extent of foreign investment. For example, the district of Buji in Shenzhen city possesses one of the most successful wholesale markets. Since its establishment in 1989, the market has expanded due to the development of agricultural activities within the Pearl River Delta, the presence of a huge urban population, a well-developed transport network, preferential policies enjoyed by Shenzhen's special economic zone, and the strong support of the local government. Another good illustration of locality effects is Shandong province's wholesale vegetable market and Shaoxing city's specialised textile market. The former is supported by the area's specialisation on high-yielding vegetable production; and the latter is bolstered by collective-owned enterprises and private dealers. In turn, these various patterns of development, or models, affect both local and national contexts and hence spatial contingency. In China, this is further complicated by the widespread propagation of purportedly successful models. Good examples are the national movement of 'learning agricultural practice from Dazhai' (*nongye xue Dazhai*) in the 1960s, and promotion of the household responsibility system during the early 1980s. Both were local experiences which influenced central policy and led to their implementation throughout China.

The above discussion highlights the importance of establishing connections between rural market development and national and local contexts. This is particularly true when scholars have to cope with China's extraordinary scale and extreme local diversity. Concepts, such as spatial contingency effects, spatial boundary effects and locality effects, which comprise the 'context' of particular places, are useful tools for this purpose. The former refers to the changes in the country's political economy with the shift from a command to a market economy during the reform era. Distinct characteristics include the decentralisation of economic decision-making power, the blending of local government administrative and entrepreneurial roles, and the rise of an administrative-zone economy. The latter – the local context – refers to the socio-economic, political, historical and cultural structures of a specific place which are demonstrated in the following chapters.

Themes of the study and research questions

Four key themes covering all major research issues in both theoretical and empirical aspects are identified in a conceptual framework to guide the direction of research. They are (a) local contexts; (b) rural market changes; (c) interactions between peasants and rural markets, and (d) the roles of local government.

(a) Local contexts

The establishment of local contexts is through the identification of political, economic and social characteristics at national and local levels. The first focuses on China's communist background, and political and economic changes during the reform era. The second concentrates attention on the economic, social, historical and environmental settings of the study areas. It covers (a) population, its distribution and size; (b) living standards; (c) overall economic performance and structure; (d) non-agricultural activities, particularly rural industrialisation; (e) agricultural development; (f) transport routes and infrastructure; and (g) extent of foreign investment. Particular interest is centred on factors which have great influence on rural market development such as rural industrialisation and transport routes.

(b) Rural market changes

Rural market development, as noted, is not regarded as a separate process. Thus, attention is focused on how the local political, economic and social settings contribute to rural market shifts. This contribution to change will be examined by analysing variations in the number of markets, their locations, distribution, hierarchy, market days and their functions at the county level. Special attention is paid to factors that affect market location. The study not only exposes how rural areas have been organised during the reform era, but also demonstrates how the present pattern is different from that described by Skinner (1964, 1965a).

(c) Interactions between peasants and rural markets

The aim of concentrating on peasant-market interactions is to show how the changing patterns of rural markets influence their interrelationship with their customers. Although peasant-market interactions are represented by flows of labour, money, vehicles and commodities (Unwin, 1989), attention is focused on (a) movement of people to particular markets, and (b) commodity flows between peasants and rural markets. This is because reliable information on the other topics is difficult to find.

Movements of people are examined by observing the marketing and shopping patterns of local residents. The former is investigated by considering commodity flows between producers (i.e. peasants) and rural markets. The aim is to study (a) how different channels are used to market crops; and (b) how administrative parameters affect market activities.

This examination of commodity flows is investigated by reviewing the study area's agricultural activities which are regarded as the local causal effects. After identifying the nature and productivity of major crops, marketing patterns of different agricultural products are explored, but only one or two major commodities are analysed in detail. Shopping patterns are then investigated by noting the choice of rural markets, frequency of visits, and mode of transport used.

(d) Roles of local government

The influence of administrative parameters is also examined through discussions with local government authorities. Dialogue is concerned with how authorities carry out central government policies and formulate local policies related to market development. The special role of local governments in managing and participating in market activities is also investigated: county-level and town-level authorities are of particular interest. Attention is paid to the management of rural markets, their administrative hierarchies, and taxes and payments imposed on trading.

These four themes address China's rural market development at two spatial levels: national and local (i.e. county). At the national level, China's changing political economy and its impact on rural market development is investigated. At the county level, the local context of a place and overall market development, locations and hierarchies are addressed. A certain number of villages are selected for detailed study, and the interactions between peasants (both as producers and consumers) and rural markets, and the roles of local government are considered. This study adopts a different approach from Skinner, who investigated the topic by comparing the patterns of both economic and administrative central places. Case studies are used for illustrative purposes in an attempt to show linkages between rural market development and a specific area's local setting, and how such connections generate spatial variations. Through this framework, the underlying causes that contribute to the pattern of current market development are addressed to provide a picture of how the countryside has been reorganised during the reform era. New complexities have been exposed. Accordingly, this thesis has revisited Skinner's (1964, 1965a, 1965b, 1985a, 1985b) classical work, particularly his hypotheses on transport improvement and marketing development, administrative and economic systems, and the logic of rural market distribution and system operation.

Deqing and Dongguan: The study area

This study focuses on Deqing and Dongguan with major attention being given to the former. The choice of these two places was determined by a desire to compare poverty and prosperity, and primitive and advanced rural market systems. Deqing typifies the non-Delta face of Guangdong Province with its limited amount of industrialisation and urbanisation, coexistence of old and new marketing systems and relatively poor development of transport facilities. Economic development has been slow and steps towards reform have been conservative. In contrast, Dongguan is well known for its rapid growth and progressive strides in economic reform. Its economic performance has led not only the Guangdong Province, but also the country. The choice of Dongguan shows how rural markets develop in an open and relatively advanced economy within a very different local context.

Apart from location, another motive for choosing Deqing and Dongguan is the availability of information on their market towns. *Deqing County Gazetteer (fangzhi)* provides a surprisingly detailed description of market days and their total trading value within its administrative boundaries. The accessibility of these sources demonstrates that records are not only available but also well-organised. This information suggests that rural markets still play an important role in their economies. Moreover, county records also provide valuable information in establishing the study area's local context. Although Dongguan has not published its accounts in the county gazette, specific materials on its commercial, agricultural and trading activities sketch the picture of Dongguan's market development. Further, Dongguan's rapid development in the reform period has attracted detailed studies of its socio-economic changes. Moreover, the availability of *Dongguan Statistical Yearbook* also provides further detailed information between 1949 and the mid-1980s.

Another reason for choosing Deqing is that several road projects are under construction which will link up several major towns. This development provides an opportunity to study local expectations of new roads, particularly their likely influence on marketing activities. Although these projects were not completed during the study period, they provide a good benchmark for future comparative studies. Finally, the author's Cantonese speaking background and well-established connections through Deqing and Dongguan have facilitated research.

Both Deqing and Dongguan are located in Guangdong Province, one of China's most prosperous areas (Figure 1.1). Although Deqing and Dongguan are similar in size, 2,257 km and 2,465 km respectively, there are remarkable differences in topography, administrative establishment, population numbers and levels of development. Deqing county is located on the north-western border of Guangdong Province and has long been regarded as one of the 50 most hilly or mountainous (*shan qu*) counties in Guangdong province. These counties have four common characteristics: (a) over 70 per cent of the county's territory is hilly; (b) an interior location; (c) poor development of transport and communication facilities; and (d) a closed or semi-closed economy. In contrast, Dongguan is located on the alluvial plain of the Pearl River Delta, where abundant arable land is available and is served by a well developed transport system. Thus, Deqing and Dongguan typify the contrasting landscapes which are described in Skinner's (1964) model of rural market structure.

Figure 1.1 Location of Deqing and Dongguan in Guangdong Province

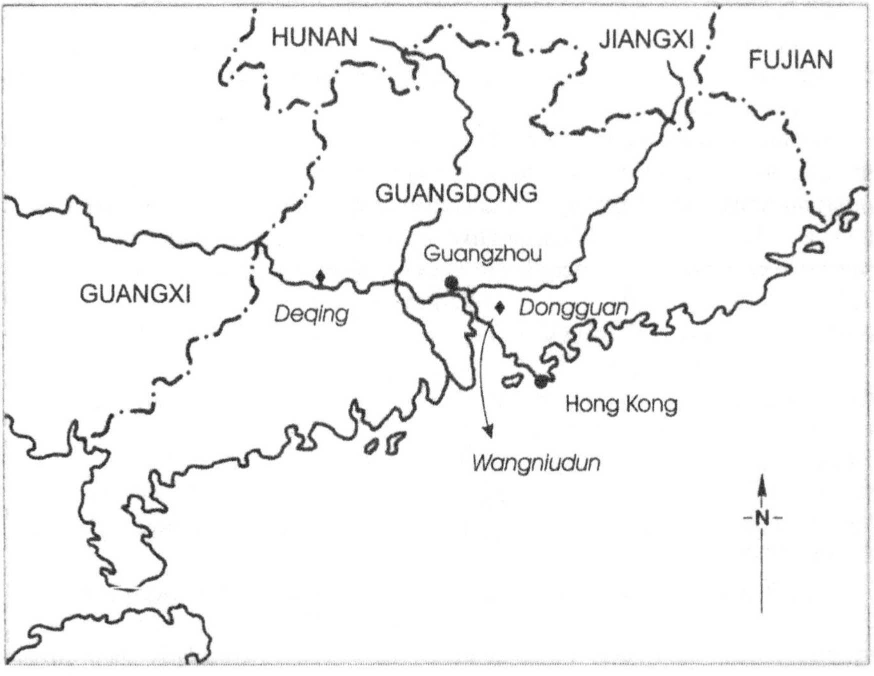

Variations in topography are not the only differences between Deqing and Dongguan; there are also contrasts in their administrative structure. Deqing is subordinate to the prefecture-level city of Zhaoqing whereas Dongguan is a prefecture-level city in its own right. Before 1985, Dongguan was a county like Deqing. Recognition of Dongguan's rapid growth led to its promotion to a county-level city in 1985 and then to a prefecture-level city in 1988.[7] Promotion in the administrative hierarchy enables Dongguan to have more control over local development projects and greater flexibility in the way it implements central government policies.

Table 1.1 Deqing and Dongguan, 2000

	Deqing	Dongguan
Area (km^2)	2,257	2,465
Population	348,440	4,073,300*
GDP (in billion *yuan*)	2.68	49.27**
Gross value of agricultural output (in billion *yuan*)	1.06	2.73
Gross value of industrial output (in billion *yuan*)	5.00	89.84
Total foreign capital used (in million USD)	13.8	1761
Per capita peasant income (in *yuan*)	3,717	6,731

Note: * including migrant workers; ** current prices.
Source: DQSB (2002), DGSB (2001).

There are also marked differences in population size and levels of economic development between Deqing and Dongguan. In 2000, the former had a population of 348,440, of whom over 80 per cent were rural (DQSB, 2002, p.241). At the same date, the population in the latter was 4,073,300 – almost twelve times larger than Deqing – but only 38 per cent of its population was classified as rural (DGSB, 2001, p.219). While Deqing organised 15 designated towns, Dongguan controlled

[7] The Chinese administrative status of '*shi*' (city) is misleading in English. Much of Dongguan remains an area of villages with agricultural activities. Accordingly, Dongguan is still referred as a rural area.

33 designated towns. In 2000, Deqing's gross domestic product (GDP) was 2.68 billion *yuan*, which ranked the county last among the six counties in Zhaoqing city and forty-seven out of 79 counties in Guangdong Province as a whole (GDSB, 2001, pp.659-660). Dongguan's GDP was 49.27 billion *yuan*, eighteen times that of Deqing, ranking it third out of all administrative zones in Guangdong Province and eleventh among the 21 prefecture-level cities (DGSB, 2001, p.220; GDSB, 2001, pp.659-660).

Given their different levels of economic development in Deqing and Dongguan, the lives of people have varied markedly. Many are dubious about Deqing's government propaganda regarding their improved well-being. In 1998, the monthly salary of a government employee averaged 700 *yuan* in Deqing. For those in joint ventures, the monthly salary was 1,000 to 1,500 *yuan*. They accounted for less than 5 per cent of the county's working population. Most people, in fact, are small farmers who cultivate several *mu* of land to support their livelihood; only a few engaged in specialised farming, and the level of specialisation in local markets is still minimal. Their tiny income suggests low levels of consumption in rural markets.

In sharp contrast to Deqing, few people have questioned the well-off status (*xiao kang shui ping*) granted to Dongguan in 1995.[8] Indeed, the lives of peasants in Dongguan have improved significantly during the reform era. Many have rebuilt their houses in concrete and added one or two storeys. Colour televisions, refrigerators and washing machines are now found in their homes. Some wealthy families also have air conditioning and satellite television. Although bicycles have been widely used, motorcycles are equally popular. According to official figures, the average annual income per capita of peasants was 6,731 *yuan* in 2000, which compared very favourably with the provincial average of 3,654 *yuan* (DGSB, 2001, p.222; GDSB, 2001, p.287). This larger per capita income suggests people have more money to spend in rural markets.

Most Deqing peasants have poor living standards. Despite the recent construction of new houses in rural areas, many still live in old mud-brick houses with tile-roofs which require frequent maintenance. This is particularly the case in hilly areas. Houses are equipped with old and simple furniture. Usually, living rooms are decorated with posters of Chinese revolutionary leaders or Hong Kong

[8] The well-off status is a comprehensive index that is determined by 16 indicators, such as per capita income, gini coefficient, life span, education level, per capita energy absorption, expenditure on clothing, and others. Amongst these indicators, income per capita is the most crucial one. According to Fan (1996), the rural and national standards of income per capita are 1,100 and 2,400 *yuan*, respectively. For details of all indicators and the estimation method, please refer to the SSB-Well-off Study Group (1992).

movie stars. Although electricity is available, people seldom turn on lights and fans because they cannot afford the high electricity bills. Few have television sets and colour television is even rarer. Telephones, washing machines and refrigerators are regarded as luxury goods and rarely found.

Many peasants in Dongguan are now divorced from agriculture. Some are working in local, publicly-owned factories and others have their own small businesses. Those still involved in agriculture are specialist farmers. Only a minor proportion of peasants still exist in Dongguan. The robust village-owned enterprises have not only absorbed surplus labour, but have also reduced tax burdens. Unlike Deqing, all taxes are paid in cash by village enterprises. Consequently, peasants no longer have to depend on cultivation and farm income for their survival.

Conversely, peasants in Deqing still depend on farming for their livelihood. As in most small-scale rural economies, all family members are involved in cultivation. School holidays are co-ordinated with the farming schedule. In Deqing, wet rice is the major crop and the sub-tropical climate allows double cropping. However, taxes account for half of the yield: essentially, one crop is produced for the tax collector. The remaining products have to be consumed by peasants and are not for sale. Although sidelines, such as vegetables and pigs, supplement their income, the increased production costs have wiped out profits and incentives. As those surpluses sold in the market produce only a small return, sideline agricultural products are largely used by the producers.

During the slack season, there are various ways to supplement their incomes from a range of *ad hoc* activities. Fish are caught in the river and wild boars or hares in the forest, and these are sold at local markets. Sometimes people will go to neighbouring villages to see if there are any temporary jobs available. As construction workers, peasants can earn up to 20 *yuan* per day. However, wage arrears are common and 'white slips' (*bai tiao*) – debt bills – are often given instead of money. The worst case scenario is that they will have worked all day but leave with nothing.

Like most areas of China, many households in Deqing have family members who are migrant workers. Discussions with peasants suggest that very few of them are lucky enough to have permanent jobs. Many are 'odd-job workers', who carry out a variety of tasks such as labouring, bricklaying and carpentry.[9] In many cases, remittances are much lower than anticipated. Some migrant labours have even asked their family for financial assistance. Still, there are some hard-workers who

---

[9] Odd-job workers are regarded as the 'three withouts' (*sanwu*) population, who are without fixed jobs, without fixed abodes and without valid documentation.

have saved enough money to build themselves a concrete house within several years. Since 1998, numerous peasants have not wanted to be migrant workers. Some of them had been cheated by employers and so discouraged from further work. Others said that it was too difficult to find a temporary job in an economy that was not thriving and they preferred to stay at home. Whether this phenomenon was the result of the Asian economic crisis was difficult to verify. Urban unemployment made it more difficult for migrant workers to find work.

Attention in Deqing is focused on the neighbouring towns of Huilong and Guanxu and some districts of Jiushi. Huilong is situated at the western border of Deqing. Since the town is dominated by hills, the population is relatively sparsely distributed. Road transport is poorly developed and there is only one paved road between Huilong and the other towns. Some villages are very remote and are connected to the outside world only by small footpaths. These characteristics explain why Huilong has been regarded as a 'corner' of Deqing. In 2000, the town had a population of 20,526 and gross rural production value (GRPV) of 199 million *yuan*, ranking it eighth and thirteenth respectively among Deqing's 15 towns (DQSB, 2002, pp.241, 256).

Guanxu, in contrast, is located in the mid-west of the county, where flatland and the population are concentrated. It was the sixth largest town in Deqing with a population of 24,088 in 2000 (DQSB, 2002, p.41). Guanxu has been one of Deqing's major grain-production areas, and yields have always topped all towns in the county. Guanxu rural market is one of the busiest. It is known as the 'home of officials' because many county government bureaucrats are from Guanxu.

Jiushi is situated along the Xijiang River on the eastern edge of Deqing. In 2000, the population in the town was 29,337, ranking it fourth among the county's 15 towns (DQSB, 2002, p.241). It has Deqing's major silkworm cocoon area, producing 153 tonnes in 1998 (DQSB, 2002, p.326). Jiushi was regarded as another 'corner' of Deqing county before a new road was opened during the 1990s. The transport improvements have significantly increased the town's accessibility and this is a key factor in choosing the town as a case study.

In Dongguan, attention is concentrated on Wangniudun, which is one of the 33 designated towns, situated 10 km on the south-west of the city government seat. The 30km^2 town has experienced tremendous changes during the reform period. A huge amount of farmland has changed to industrial use. Township and village industries have flourished and the incongruous scene of factories surrounded by paddy fields is all too familiar. By 2000, the town had more than 200 factories, including 88 foreign-invested firms producing exporting commodities. In 2000, these factories, together with Wangniudun's agricultural sector, contributed 364 million *yuan* of GDP (DGSB, 2001, p.236).

Précis of the study

Chapter 2 reviews Skinner's (1964, 1965a, 1965b, 1985a, 1985b) classic thesis on China's rural market development. Skinner's (1977, 1994) regional approach is also presented. His major arguments on rural market growth, intensification, modernisation and its development during different time periods are highlighted. Also, the literature on China's post-reform rural market development is reviewed. This establishes the conceptual framework of the study.

Chapter 3 investigates China's changing political economy during the reform era and its impact on rural market development. This is done by identifying spatial contingent effects associated with China's national level. Particular attention is placed on the decentralisation of power and the formation of an administrative zone economy. By examining the state's role in market development, this chapter argues that the traditional market system – seen by Skinner (1964) as a 'natural' system based on market forces – has not been revived during the reform era.

Chapter 4 analyses rural market development at a local level. Deqing and Dongguan are used as case studies. Particular attention is given to spatial boundary effects on different localities and their impact on rural market distribution, hierarchy and functions. It shows how variations in local socio-economic, political and environmental structures have generated different levels of market development in Deqing and Dongguan. This chapter also examines the relationship between population and market growth in the two places. Comparisons are made not only between Deqing and Dongguan, but also between them and Skinner's (1964, 1965a) hypotheses and models.

Chapter 5 further analyses Deqing and Dongguan's marketing pattern by examining locality effects and their impacts on peasant-market interactions. Specific attention is given to economic factors such as the level of self-sufficiency, income and transport improvement. By examining these elements and contrasting them with Skinner's equation of market evolution, this chapter demonstrates the more complex situation that has developed during the reform era. The findings in this chapter also elaborate and supplement Skinner's mid-1960s thesis.

Chapter 6 extends discussions by focusing on administrative aspects. Administrative principles, government regulations, toll stations and administrative boundaries and their impact on Deqing's and Dongguan's rural marketing activities are examined. Canvassing these factors demonstrates that Skinner's (1985b) policy-cycle theory is not sufficient in itself to explain changes in the political economy of the reform era.

Chapter 7 elaborates the findings of previous chapters and examines whether these variations are sufficiently explained by Skinner's (1977, 1994) regional framework; his core-periphery concepts and classification of macroregions are discussed. While demonstrating the inadequacy of these concepts, this chapter suggests a broader urban-rural perspective to understanding development.

The last chapter – Chapter 8 – concludes this study by discussing the extent to which Skinner's thesis applies to post-reform China. Future developments of the country's rural market and its reform are reviewed. Finally, future directions for research are outlined.

Chapter 2

China's Rural Market District: The Central Place Theory and Skinner

Administrative parameters have exerted a strong influence on China's regional economic development. As a centrally planned economy, the role of the state and its policy implications for regional economic development have always been a major concern. During the pre-reform period, a rigid governmental system was implemented and China was thus divided into different 'areas' (*kuai kuai*) by administrative boundaries. Urban and rural areas were organised by different administrative units. Lateral connections were prohibited, and a social system was established which separated China's population into urban and rural classes. This system determined an individual's educational chances, work opportunities, marriage prospects and other social benefits.

In 1978, economic reform altered most of these radical policies. Whether China's social economic development has broken away from administrative influence, however, is debatable. Some studies show that economic development has integrated different regions (Marton, 2000; Lin 2001a, 2001b). Others demonstrate that the administrative boundaries are acting as 'invisible walls' and have constrained economic integration (Huang, 1993; Liu and Shu, 1996; Tang and Chung, 2000, 2002). In fact, economic reform has not changed the practice of using administrative measures to manipulate economic development. Cities are still being regarded as administrative organisations rather than central places within a purely economic system (Chung and Tang, 1997, p.3).

The importance of administrative boundaries in China is forcefully illustrated by Skinner's (1964, 1965a, 1965b, 1976, 1977c, 1985a, 1985b) studies on China's rural market system. His research provides an historical framework for understanding China's marketing structure and its relationship with the administrative system from the late imperial period to Mao Zedong's China. Before discussing and reviewing Skinner's work on critical periods in the development of market districts, it is important to review his development of earlier explorations on central place theory by Christaller and Lösch.

Central place theory: Skinner, Christaller and Lösch

G. William Skinner graduated with distinction in Far East Studies at Cornell University in 1947. Subsequently, he undertook a Ph.D. in anthropology on 'A Study of Chinese Community Leadership in Bangkok together with an Historical

Survey of Chinese Society in Thailand' (Skinner, 1954). Since then, he has extended his interest in cultural anthropology, historical anthropology and comparative agrarian society in Asia, with particular reference to Chinese communities overseas. Field research has been conducted in Borneo, China, Japan, Java and Thailand. During the late 1950s, Skinner published his findings on the Chinese overseas in *Chinese Society in Thailand (1957)*, and *Leadership and Power in the Chinese Community of Thailand (1958)* which were based on his Ph.D. thesis.

Since the late 1950s, Skinner has concentrated on the People's Republic of China. He has written about China's rural marketing system (Skinner, 1964, 1965a, 1965b), Chinese society (Skinner, 1951, 1971, 1973a, 1973b, 1973c), China's urban and regional systems (Elvin and Skinner, 1974; Skinner, 1976, 1977a, 1977b, 1977c), and China's history (Skinner, 1982, 1985c). In 1994, he presented a theoretical framework for comprehending China's regional differentiation (Skinner, 1994).

Skinner's (1964, 1965a, 1965b) most significant contribution, however, was arguably his series of papers on China's rural marketing and macro trading systems. As these papers employed central place theory to study the distribution of China's rural markets and their relationship to the administrative system, the discussion is centred on Skinner's debt to Walter Christaller (1933 trans 1966) and August Lösch (1940 trans 1954), who were originally responsible for developing the underlying concepts. Undoubtedly, their formulations appealed to Skinner's meticulous nature (Unger, pers. comm.).

Skinner elaborated Christaller's 1930s work on economic central places to seek "a general explanation for the sizes, numbers and distribution of towns" (Christaller 1933 trans 1966, p.2). Christaller argued that the demand threshold for the supply of high-order goods determined its marketing area, which was usually hexagonal in shape. The marketing area also determined the commodity's minimum price. This process was the same for second-order and third-order goods which produced different levels of centres. Each central place served an optimal pattern of settlement which was hexagonal in form. Small hexagons (i.e. low-order centres) were nested in the larger one (Figure 2.1).

Skinner also drew on Lösch's (1940 trans 1954) location theory of scattered farms, despite its assumption of pure competition and its 'bottom-up' approach. Under the assumption of pure competition, suppliers would locate themselves in such a way as to prevent excess profits being earned. The 'bottom-up' approach assumed that small centres would be established first, while larger ones would stem from demand intensification. Further, Skinner elaborated Lösch's (1940 trans 1954) notion on the indissoluble connection between transport routes and the spatial patterns of central places. He stressed the significance of road routes in shaping rural market development and its service areas. Obviously, his perspective reflected the booming interest in transport and development studies during the late

1950s and 1960s (Owen, 1959; Green, 1958; Berry, 1959; Garrison, Berry, Marble, Nystuen and Morrill, 1959; Fulton, 1969).

Figure 2.1 Regular central place hierarchy

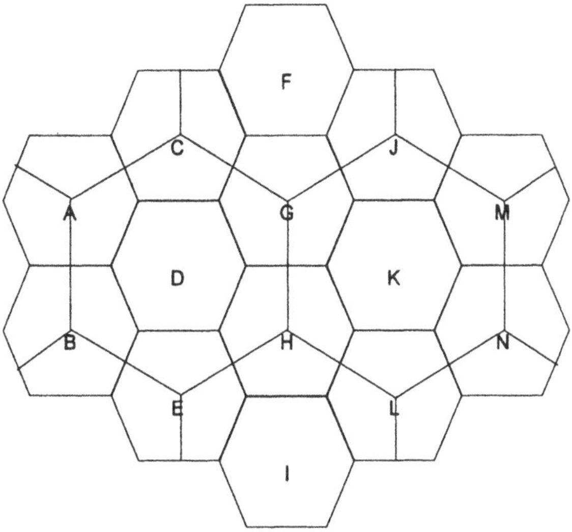

Note: Small and large hexagons represent central places at low and high levels respectively.

Skinner used the contributions of Christaller and Lösch "to explain the distribution of retail market centres, the purest examples of which are periodic 'peasant' marketplaces" (Smith, 1974, p.171). Several studies applied his hypothesis of marketing evolution and the relationship between marketing and social systems and produced fruitful results (Smith, 1976a, 1976b).

Unlike Christaller who began with high-order goods and urban centres, Skinner adopted a 'bottom up' approach to develop his central place arguments. Initially, his analysis centred on small, rural periodic markets. Population growth raised the demand for agricultural commodities and the level of specialisation. In addition to transport development, markets experienced a cycle of intensification which eventually led to the formation of urban centres. Accordingly, urban centres, in Skinner's thesis, are formed as a result of rural intensification. By adopting this perspective, Skinner has overcome the most trenchant criticism of central place theory – its static nature. Simultaneously, he developed a "complete model of central-place evolution" which Christaller and Lösch did not work out for themselves (Smith, 1976a, p.46). Skinner's evolutionary concept is regarded as an important anthropological contribution to geographical studies. Besides, providing a systematic framework for anthropologists to study marketing systems, his peasant marketing model has also contributed to debates among geographers (Marshall, 1964; Berry, 1967; Berry and Horton, 1970).

In recasting central place theory, Skinner's thesis also modified some of its unrealistic assumptions. His model overcomes the need for (a) homogeneous topography; (b) uniform demand density and purchasing power; (c) equal transport facilities and services; and (d) indefinite extension of the same development pattern in all directions (Skinner, 1977c, p.281). Instead, he has proposed a range of systemic variations (Skinner, 1977c, pp.283-285). Demand density varies as a function of population density and distance, and is reflected in transport costs. In sum, Skinner is concerned with the contribution of topography and socio-economic parameters.

Skinner's use of central place theory to articulate market systems also attracted the attention of anthropologists. Heated discussions took place on this topic during the 1970s, particularly among Stanford anthropologists. Indeed, central place theory, according to Smith (1974, pp.167-168), provided Skinner with a systematic perspective for understanding the marketing behaviour of peasants which was much broader than the usual anthropological view of peasant communities. Subsequently, five aspects of the central place theory have attracted the attention of anthropologists: (a) the delimitation of a village or city as a research unit; (b) the measurement procedures for identifying development at a regional level; (c) the framework for describing trade articulation; (d) the framework for understanding peasant communities; and (e) the ability to tackle economic and social relationships (Smith, 1974, p.170). As a result, anthropologists "need less modification for the analysis of agrarian marketing", unlike economists and geographers who have to modify the theory's simple assumptions to understand economic activities complicated by modern transport and localised production (Smith, 1974, p.169). This finding demonstrated central place theory's potential for further elaboration and application in studying China's marketing systems.

Market districts pre-1949

The pre-1949 market districts in rural China were regarded by Skinner (1964) as a traditional system, comprising three levels: standard, intermediate and central. The 'standard market' was the core, as it was "the starting point for the upward flow of agricultural products and craft items into higher reaches of the marketing system, and also the termination of the downward flow of imported items destined for peasant consumption" (Skinner, 1964, p.6). It was the first place for peasants to sell their products and to purchase daily necessities. In contrast, the 'central market' was the upper reach of the market hierarchy and located at a key urban node in the transport network. It had important wholesaling functions and served as a high-level market for exchanging local and imported commodities. As regards the 'intermediate market', Skinner (1964, p.7) does not define it precisely but regards it as having "an intermediate position in the vertical flow of goods and services both ways". The settlements in these market towns were denoted as either a 'standard market town', 'central market town' or 'intermediate market town' respectively.

These three levels of market towns and cities reflected another order of central places in Imperial China. Unlike the marketing system, which was formed by 'natural' economic activities, this was shaped by administrative and political transactions and regarded by Skinner (1964, 1976, 1977c) as an 'official' or 'artificial' system. This highlighted an important issue in Skinner's research: the relationship between the market and administrative hierarchies. Skinner (1977c) emphasised the primacy of the market structure characterised by central economic functions. This is because economic activities, notably resource allocation, profits and extraction, raised the need for governmental regulation and control. Thus, economic central places were the logical sites for administrative institutions (Skinner, 1977c, p.276). Accordingly, when considering the relationships between the market and administrative systems, Skinner (1964, p.9) stated that:

> The extent to which the two hierarchical series of administrative and economic centers overlap or coincide can be determined......only through an analysis of marketing structures in a given region.

Given the primitive transport linkages and a 'closed' agricultural system, the traditional market structure was characterised by its periodicity and the participation of mobile traders. Periodicity reflects China's extraordinary scale and local cultural variations. Different parts of China followed their own scheduling arrangements determined by the lunar decade (ten days), independent decade and independent duodenary (twelve days) (see Skinner, 1964, pp.10-16 for details). Generally though, market cycles followed the lunar calendar used by Chinese peasants for thousands of years. A two-per-lunar decade schedule was most common (Skinner, 1964, p.14). While neighbouring standard markets often had the same schedules, other levels had different timetables. This prevented standard market days from conflicting with intermediate market days. These arrangements allowed vendors and villagers to visit markets at different levels without any time clashes.

As argued by Skinner (1964), China's pre-modern rural market system could be explained by Christaller's (1933 trans 1966) regular hexagonal pattern. Based on data from local gazettes and records, Skinner (1964, p.18) maintained that a standard market generally had a service area of about 18 villages. After considering China's topographical differences, he developed two models to explain market patterns and their spatial expression in both plains and mountainous areas (Skinner, 1964, 1965a) (Figure 2.2).

Figure 2.2 Skinner's models of different topology

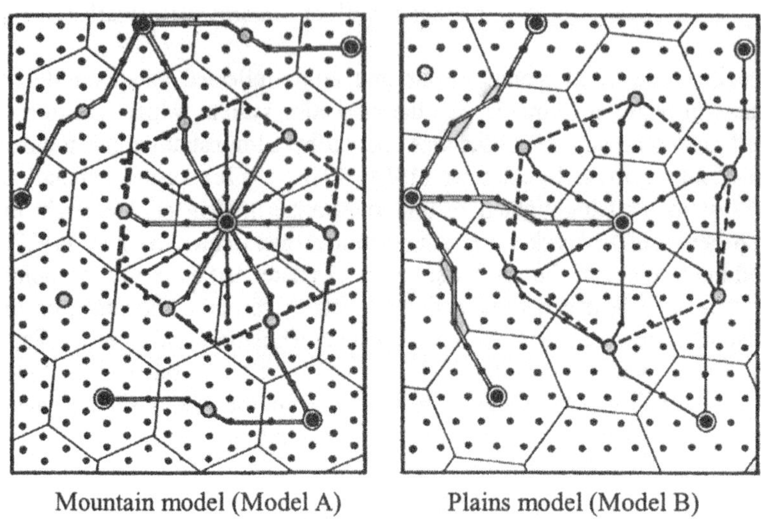

Mountain model (Model A) Plains model (Model B)

Source: Skinner (1965a, p.203).

The differences between these models were expressed by reference to Sichuan province. In the plains of Sichuan province, typified by abundant arable land and high accessibility, each standard market town, on average, depended on three intermediate markets. Conversely, each standard market town in Sichuan's mountainous areas, which had fewer possible transport routes and limited arable land, depended on two higher-level market towns (Skinner, 1964, p.21). The marketing territories in mountainous areas were also larger than in the plains, as the population was more scattered. According to Skinner (1965a, p.205), a complete intermediate market covered an average area of 235 km^2 in mountainous areas, while only 105 km^2 in the plains.

Skinner (1965a, p.206-211) suggested that the local market system will pass through an intensification cycle during its evolutionary process, despite the differences in patterns between the plains and mountain models. This process is brought about by a rise in rural population and an increase in the degree of household participation in the market. The intensification cycle is triggered by an increase in market size measured by its physical area or number of participants. Any expansion is followed by a rise in the number of trading days and eventually an increase in the number of markets. Initially, according to Skinner (1965a, p.206), village-to-market ratio might rise to 40 to 50 villages per market in mountainous areas where the transport network was poorly developed. An increase in the number of market days was accompanied by a rise in higher-level functions, such as wholesaling, and the eventual setting up of new standard markets. Once these were established, the village-to-market ratio reverted to the normal level (i.e. one to eighteen).

The intensification cycle was accompanied by a process of modernisation which was characterised by greater commercialisation. It was reflected in a rise in the number of permanent traders and facilities, and in the degree and scope of specialisation at the expense of peasant self-sufficiency (Skinner, 1964, pp.211-213). However, this did not necessarily entail modernisation, as Skinner (1964, p.216) argues:

> True modernisation occurs only when a modern transport network is developed within an already commercialised central marketing system.....By contrast, commercialisation without intrasystem transport improvement amounts to a kind of false modernisation.

The development of a modern transport system also has a positive effect by speeding up the intensification cycle. It is regarded as a dominant factor shaping both rural market intensification and modernisation. Generally, the market towns which are connected to the central markets by new transportation routes are the first to be modernised. Conversely, those bypassed by the new facilities are unlikely to be modernised and may even cease to exist. According to Skinner (1965a, p.220):

> [I]n the ideal case all traditional standard markets will have died while all traditional higher-level market towns will have been transformed into modern trading centres. At this point, the basic economic unit may be designated a modern trading system as opposed to a traditional marketing system.

By drawing examples from different parts of China, such as Shanghai, Nanjing and Ningbo, Skinner (1965a, pp.221-226) concluded that 'true modernisation' had progressed very little before 1949. Although transport improvement and agricultural commercialisation had permeated east China's entire agrarian economy, modernisation of markets had only:

> proceeded to the point where standard markets were extinct......often only to the extent of [a] city's own intermediate marketing system......sometimes to the extent of [a] city's own central market system (Skinner, 1965a, p.222).

The influence of the administrative system on that of the market structure was rare before 1949.

The marketing hierarchy and regional approach

China's rural market system is regarded by Skinner (1964, p.32) as the basic locus of a peasant community. It is connected to similar local market structures horizontally and aligned to a higher level system vertically. Skinner (1977b) identifies eight levels in the regional hierarchy during Imperial China. In ascending order, these levels were: (a) standard market; (b) intermediate market; (c) central market; (d) local city; (e) greater city; (f) regional city; (g) regional metropolis; and (h) central metropolis. Low-level systems were "nested within [higher level] systems, and so on in the manner suggested by central-place models" (Skinner, 1977c, p.286). Each system's hinterland was defined by population density and transport networks and was not constrained by administrative boundaries. These systems finally made up eight macroregions in China, they are: North China, Northwest China, Upper Yangtze (*yangzi*), Middle Yangtze, Lower Yangtze, Southeast Coast, Lingnan (where I conducted fieldwork research) and Yun-kwei (*yungui*). Each has a core, centring on one or more major commercial cities, and peripheries comprising far less commercialised outer hinterlands.

Market districts after 1949

Local rural market systems underwent marked fluctuations after the establishment of the People's Republic of China (PRC) in 1949. As demonstrated by Skinner (1985b), these were caused by a persistent cycle of policy changes comprising liberal, radical and crisis phases. The cycle began when policies favouring market development were implemented during a liberal period. When the economic situation improved and enthusiasm grew, policies were radicalised. Invariably, these reforms triggered a crisis stage when production declined and peasant alienation increased. To tackle the impending crisis, liberal policies had to be re-implemented, and another cycle started. Between 1949 and 1977, Skinner (1985b, p.397) identified 11 cycles during which shifts in policies accounted for the recurrent repression and revival of China's rural markets.

The first few years after the establishment of the PRC were regarded as a 'liberal period'. No significant drawbacks were experienced by the traditional marketing system despite: (a) the establishment of state agents, such as state trading companies and supply and marketing co-operative, to plan and socialise rural trade; (b) the implementation of 'unified purchase, unified sale' – state monopoly of the trading of all agricultural products; and (c) the dominance of state-owned enterprises in the wholesale market which led to the schedules for various types of markets being rigidly retained (Skinner, 1965b). Similar meeting patterns persisted and mobile traders maintained their standard market circuits. Intermediate markets and central markets continued their respective wholesale functions, though most firms were taken over by state companies (Skinner, 1965b, p.364). Although a more efficient transport system linked up industrial centres and central market towns, "efficiency within intermediate and standard marketing system was little improved" (Skinner, 1965b, pp.366-367). As argued by Skinner (1985b, p.395), it was during liberal phases "that China most closely resembles a traditional agrarian society going about its usual peasant business". In other words, rural markets during the early years of PRC remained the same as those during the traditional period.

Instead of growth and development, the market system suffered when hard-line policies were implemented during radical phases such as the Great Leap Forward (1958-1961). People's Communes were established to organise agricultural production and consumption in a single department under the direct guidance of state policies and trading organisations. Simultaneously, a series of anti-market measures was instituted such as the elimination of rural free markets, a prohibition on peasants marketing their surplus, and the prevention of peasants having private plots. Consequently, periodic markets were closed down, rural marketing activities were halted and commodity exchanges and distribution in most parts of China experienced "quickly induced near paralysis" (Skinner, 1965b, p.372).

The attempt to use administrative measures to reorganise marketing activities was a total failure. Skinner (1965b) maintained that the integration of the People's Communes and the market structure failed because the new administrative unit was not aligned with the natural rural trading order. This was attributed to the function and size of communes. The reorganised marketing system together with the new administrative model, the People's Commune, had destroyed the natural mechanism of the marketing practice. As argued by Skinner (1964, p.31), "marketing structures, unlike administrative structures, take the form of interlocking networks". In other words, "whereas administrative units are discrete throughout the hierarchy, each lower-level unit belonging to only one unit at each ascending level, marketing systems are indiscrete at all levels except that of the standard market" (Skinner, 1985b, pp.401-402). By attempting to align the market system to the administrative hierarchy, the rural commercial exchange organisation was eliminated under the People's Commune with ruinous effects.

The size of People's Communes was driven by revolutionary zeal and the desire of local cadres for personal achievement. Consequently, most communes were 'over-sized' because they exceeded the upper-limit of the standard market system. This led to an inefficient distribution of commodities and an overloading of transport and storage facilities. At the same time, the formation of the People's

Commune stimulated localism, or, in Skinner's (1965b) term, "local particularism". Conflicts occurred between the various collective brigades formed by different villages. Together, these events demonstrated that the People's Commune had interfered with the normal mechanism and operation of China's rural marketing system. Also, the new structure failed to take into account variations in the character of different villages. The result was a complete breakdown of the natural rural market and the spatial division of labour.

The influence of administrative factors on rural market development was much greater in socialist China. This stemmed from the state's attempts to use regulatory measures to manipulate economic activities. In fact, China, as a planned economy after 1949, had established a centralised administrative framework to manipulate every single aspect of the economy. Although Skinner (1985b) did not mention administrative factors in his analysis of the policy cycle, different phases in the cycle can also be interpreted as different degrees of government control. Accordingly, in the radical phase, when the state's control was tight, market development was suppressed. Conversely, rural markets revived in the liberal period when state control was less secure.

During the early 1960s, rural markets re-opened and the size of People's Communes was reduced so as to retrieve the 'natural' standard market district. The marketing system, however, developed at a slower rate. This was because the re-opened markets were under the strict control of various committees formed by local cadres and other authorities (Skinner, 1965b, p.373). As a result, the commodities, prices, participants, and marketing practices were restricted. Owing to political and social instability, the development of the rural market system during the mid-1960s and early 1970s was difficult to investigate. Repression of rural marketing for ideological reasons was the general trend during Mao's era (1949-76) which led to an overall decline in numbers. Skinner (1985b, p.405) concluded:

> The replacement of periodic marketing in rural China by an efficient modern distributional infrastructure is, of course, a perfectly reasonable policy objective. The fundamental error of the radical programme for rural commodity distribution during the Maoist era was the persistent attempt to restrict, impair or eliminate rural periodic markets before a viable modern infrastructure was in place.

State control, therefore, had repressed the development of rural markets, their evolution ceased and modernisation progressed extremely slowly throughout Mao's era.

Market changes since 1978

Economic reform since 1978 can be regarded as another liberal phase, when a "steady and methodical deradicalisation" occurred (Skinner, 1985a, p.32; 1985b, p.405). This period has witnessed the de-collectivisation of the farming system, the abolition of People's Communes, the relaxation of controls and the re-introduction of the household farming system; changes which have resulted in restimulation of peasants' incentives in agricultural production and brought positive effects to the rural market system. As argued by Skinner (1985b, p.410):

> [T]he transformation of higher-level periodic market towns into modern trading centres, described in my 1965[a] article as having begun in China after 1895 in the cores of China's most advanced regions, is under way once again. Even as traditional dynamics are playing themselves out in the reconstitution of marketing systems and their further extension into regional peripheries, modern change is at work in the cores of the more developed regions.

Whether the traditional marketing system has re-emerged in rural China is open to debate. Scholars, such as Findlay, Watson and Martin (1993) and Watson (1988, 1996), agree that rural markets have re-emerged. They have argued that reform measures, particularly price reform and the withdrawal of state interventions in commodity allocation, have stimulated their growth. According to Watson (1988) and Findlay, Watson and Martin (1993), the number of free markets doubled between 1978 and 1991, and their trading value increased from 125 million *yuan* in 1978 to 838 million in 1989. Further, Watson (1988, p.17) argues that the interlocking hierarchy of the pre-modern Chinese marketing economy re-emerged in the reform era. However, despite the dramatic increase in the number of free markets and their trading value, a recent World Bank (1994, pp.69-99) report highlights that the state still plays a surprisingly dominant role in commodity distribution. It is, therefore, debatable whether the marketing system has been fully reconstituted, given the special political and economic contexts of the reform era in which the state and market mechanisms have been juxtaposed.

Findlay, Watson and Martin (1993) state that the rural market system during the reform era has several distinctive characteristics. First, the juxtaposition of state intervention and market forces or, in general, a double-track system (*shuang gui zhi*), has emerged from the persistence of state intervention. State institutions, for instance, continue to play an important role in the agricultural marketing system. In some cases, they are both organiser and participant in the same market-place (see Watson (1988) for examples). Consequently, although state units, collective units and individuals participate and compete in the market, competition is not necessarily even. Further, the state marketing and free marketing systems not only coexist in retail markets, but also in wholesale ones.

Secondly, the emergence of long-distance and specialised markets through the development of transport networks – particularly railway and road transportation – has fallen considerably behind economic growth during the reform era (World Bank, 1994). This new direction has placed extra pressure on existing transport networks and pushed governments at all levels to improve them. In this case, the rapid growth of different types of markets has accelerated modern transport development – a causal relationship different from Skinner's hypothesis. Finally, the function of the market has changed as rural markets are no longer places where peasants merely exchange their surplus (Watson, 1988; Findlay, Watson and Martin 1993). The market is regarded as a place to raise the income of peasants, to absorb rural surplus labour, to stimulate competition among different parties, and to increase domestic demand so as to maintain overall national economic growth (Watson, 1988; *Renmin Rebao* (People's Daily), 19 October, 1998). All of these features imply the emergence of a new marketing system which is very different from the traditional one.

Another feature of China's market development during the 1980s has been the 'commodity wars' between different regions which occurred at both village, county and provincial levels. Local governments used various trade barriers, such as fees, fines, roadblocks or even force, to stop the 'export' of primary products so as to ensure the local processing of these products (World Bank, 1994, p.38). Throughout the 1980s, commodity wars, such as the 'wool war', the 'tea war' and the 'grain war', and other small-scale interregional trade conflicts, were reported in different parts of China (Watson, Findlay and Du, 1989; Forster, 1991; Zhang, Lu, Sun, Findlay and Watson, 1991; Watson and Findlay, 1992; World Bank, 1994). These observations imply market development is constrained by unstable factors brought about by reform. As long as trade barriers exist, it is misleading to conclude that an integrated market system – like the traditional system described by Skinner – has revived in China.

Skinner's (1985b) model is therefore only partly applicable in the reform era, despite his argument that the market development process described in the mid-sixties articles has been reactivated. Most of Skinner's (1964) contentions were based on the analysis of a closed rural system within a small-scale peasant economy. As noted, his assertions highlighted (a) the importance of demand density in determining the size of market systems; (b) the three functional levels of market towns; and (c) the dominant role of transport in market modernisation. In the traditional system, peasants were self-sufficient rather than dependent on pure commodity exchanges. Markets, for instance, were places they could exchange agricultural surplus, and specialisation only occurred within a very narrow range of crops. As anticipated, once the closed system was broken down by the development of industrialisation, urbanisation and inter-regional trade, a new market structure was formed with very different functions and mechanisms. Apart from transport, other factors, such as population growth, commodity specialisation and trading which played a minor role in Skinner's (1964, 1965a, 1965b) analysis, now play a more active role.

Another major concern of Skinner's study of rural markets is the impact of the administrative system. Although administrative centres were closely integrated with market towns in pre-communist China, the role played by the state was very minor (Skinner, 1985a). An extreme development in the state's role was witnessed in Mao's era when the People's Communes were established, paralysing rural marketing activities. Two key questions arise: to what extent do government regulations or administrative factors exist in the era of economic reform; and what have been their impact on rural market development?

One reason for the failure of the People's Commune was its attempt to control everything. The extent of state control in China has always captured the attention of academics and heated discussions have arisen. While some scholars, such as Naughton (1995b), have argued that economic reform has weakened the state's power, others such as Zweig (1992) and Shue (1988, 1995) have contended that state authority has remained unchanged. Economic reform has contributed to the combination of the administrative and entrepreneurial roles of local governments (Oi, 1992, 1995, 1999b; Unger and Chan, 1996). Also, administrative measures have been used to facilitate local economic growth (Blecher and Shue, 1996; Marton, 2000). Further, Chung and Tang (1997) have asserted that local governments have used administrative boundaries to protect their own interests. Two further questions are raised: what has been the impact of administrative factors on the marketing system; and is it any different from that of the Maoist era? This is analysed in subsequent chapters.

Skinner's thesis of market development has been challenged by a number of new features during the reform era. As noted, the new features include: (a) juxtaposition of the market mechanism and state intervention; (b) the growth of markets despite state intervention; (c) the changing functions of rural markets; (d) regional trade conflicts and barriers; (e) the new causal relationship between market and transport development; (f) the influx of foreign investment; and (g) the rapid development of rural industrialisation. Although some of these have been addressed by Watson (1988, 1996), and Findlay, Watson and Martin (1993), their studies are dominated by economic perspectives. Accordingly, they focus on the growth of various kinds of markets and their relationships with the state apparatus to sketch the development of the market economy in the reform era. Like most economic studies, they neglect spatial factors and their effects on market development during this period. A detailed study of the market system, its spatial distribution, hierarchy, and relationships with the administrative hierarchy has still to be undertaken.

In short, Skinner's studies provide a useful framework for understanding market development. However, the complexity and rapid changes in political economy during the reform era have created a new context and generated a different set of spatial contingencies and causal effects. Consequently, the explanatory power of Skinner's model, which was established during the 1960s, has to be re-examined through new research on China's rural marketing system, its patterns and problems during the transformation from a hard-line planned economy to a 'socialist market economy'.

Chapter 3

China's Rural Market Development Since 1979: Has the Traditional Market System been Revived?

Economic reform since 1979 has been regarded as a 'third revolution' in China's countryside (Garnaut and Ma, 1996). The abolition of the People's Communes and the re-implementation of household farming systems gave peasants more control over production and boosted agricultural output. Grain and cotton yields reached a record high in 1984. Other major crops, such as oil crops, beets and sugar cane, also recorded over 10 per cent growth between 1978 and 1984. This substantial expansion created huge surpluses and pushed the state to liberalise trading and marketing activities, which had been attacked as the 'tail of capitalism' during Mao's era.

The revival of China's rural markets in the reform era is attributed by some scholars to the unleashing of market forces following the withdrawal of state control. This argument is unconvincing because it oversimplifies what has happened in China. Economic reform should be regarded as a redistribution of political power between the state and local governments, with more control decentralised to the local level; a process which has changed the inter-relationship between state and local governments and created a new context during the reform era. Since many studies have overlooked this changing political economy, they fail to explain why the extent of market liberalisation has varied between and within different regions and why market problems have arisen in some areas and not others. Moreover, given the tradition of intervention in socialist economies, it is doubtful that a traditional system – driven by market forces – would be entirely revived.

This chapter examines China's rural market development during the reform period by employing the contextual approach and spatial concepts discussed in the pervious chapter. The changing political economy since 1979 and its impact on rural market growth and development is examined, as is the government's role in establishing rural markets, and their subsequent operation and development. This chapter argues that the current market system is different from the traditional one as described by Skinner.

The spatial contingent effects: China's changing political economy

China's economic reform, which is regarded as the most "significant global event in the final quarter of this [20th] century" (Lees, 1997, p.11), began in 1978, two years after Mao Zedong's death. It is now understood as a process rather than a single event. At that time, China's economy had been plagued by low productivity, inefficiency, and shortages for decades. Between 1950 and 1979, industrial productivity had declined by an average annual rate of 2.75 per cent (Lees, 1997, p.16). In the agricultural sector, grain production fluctuated between 250 and 282 million tonnes during the period 1971 and 1976 (SSB, 1989, pp.367-368). Initially, a series of adjustments was made to correct the distortions that prevailed under the Maoist development model, namely, an autarkic system ideologically hostile to the western world. As noted, these adjustments included the abolition of the People's Communes and implementation of the household responsibility system. At this stage, a master plan and set of objectives had not been formulated. As reflected in Deng Xiaoping's slogan *mo zhe shitou guohe* ('crossing the river by groping for stepping stones'), a clear goal – the bank of the river – was lacking.

The early economic restructuring was successful (Ash, 1996; Huang, 1998). Between 1976 and 1984, grain production increased from 282 to 407 million tonnes – an average annual growth rate of 15 per cent (SSB, 1989, p.368). By 1985, the number of rural markets had doubled. Between 1978 and 1985, market turnover value soared from 12.5 billion *yuan* to 63.2 billion *yuan* (SSB-GMSD, 1999, p.233). In the industrial sector, the annual growth rate of gross industrial output value (GVIO) also expanded from 2 per cent in 1976 to 16.3 per cent in 1984 (SSB, 1989, p.376). Significantly, economic development had not only increased productivity and incentives in rural areas, but also reshaped individual attitudes and expectations at all levels of society. In turn, these outcomes encouraged the state to extensively restructure other sectors of the economy. Although these reforms were initiated by a series of *ad hoc* adjustments, since the early 1990s measures have been introduced to resolve specific issues, particularly the need to intensify competition, increase efficiency, combat corruption, eliminate trade barriers and protect property rights.

The reforms, as discussed by Rawski (1999, p.142), have involved "an extended process and replete with interaction and feedback" between the central and local states, government authorities and civil society, enterprises and workers, and distributors and consumers. The process is dynamic, with frequent changes in the relationships between actors, and different causal mechanisms and contingent effects being generated. These are subjected to further change when new measures are introduced. Rural markets, therefore, have been developed in a complex and dynamic environment. Before addressing China's rural market development, it is imperative to have a clear understanding of the country's changing political economy.

The Chinese state during the pre-reform period was generally regarded as a 'police state', characterised by state intervention in economic, social and other realms of life through "techniques of hierarchical observation, normalisation and examination to produce governable socialist persons" (Tang, 1995, p.5). Economically, the state owned almost all productive enterprises and administered their investment, output, use of technology, distribution of commodities, appointment of personnel and the salaries of workers, through a multi-level administrative hierarchy. The supply, demand, price, and flow of commodities were concomitantly manipulated by the state. Socially, the household registration system was implemented to control population size and mobility. An individual's education, social welfare, work opportunities, marriage, allocation of grain and other daily necessities were all controlled through this system.[1] In short, every aspect of life was covered by centrally imposed policies.

Whether this 'police state' has continued to exist in the reform era is debatable. Economically, the success of rural reform, the rapid economic growth of core rural areas, and the shrinkage of the planning system have diminished the state's capacity for economic control (Naughton, 1995a, 1995b). Price control has been relaxed and distribution channels have been opened to include private agencies. Gradually, the allocation of goods has been removed from the planning authorities, and special agricultural and industrial products markets have been established. A study by the World Bank (1994, pp.xiii, 193) shows the state was responsible for less than 21 per cent of retail sales in 1991; in 1993 only 10 per cent of industrial goods were subjected to state allocation. As well as the dramatic growth of rural markets, special wholesale markets for grain, rice, sugar and poultry were established in Zhengzhou, Guangzhou, Yujian and Wuxi cities respectively (World Bank, 1994, p.102). Further, state control over personal freedom and community life has diminished (Shue, 1995). The household registration system no longer constrains an individual's education, career choices, and movements (Gu and Li, 1995; Mallee, 1995). Peasants are permitted to migrate to the cities as long as they provide their own food. University graduates are also allowed to work in rural enterprises without losing their urban residence and privileges.

Despite the above changes, some scholars argue that state control has not altered during the reform period (Shue, 1995; Huang, 1996a; Zweig, 1997; Muldavin, 1998). Examining the central-local relationship, Chung (1995, p.502) argues that decentralisation of power has not "necessarily and automatically" weakened the state's control. These scholars point to its hold over major heavy industries, intimate ties with some enterprises, control over import and export licences and quotas, and monopoly procurement of cotton, tobacco and some local,

---

[1] See Guo and Liu (1990) for a detailed study on the issue.

special agricultural products. Mastel (1997) even regards the continuation of five-year planning – a unique product of the socialist economic system – as an indicator of China's uncompromising stance against economic liberalisation. The one-child policy is a good illustration of the state's control over social matters and, politically, it still manipulates personnel appointments in key governmental positions. Thus, though state control has diminished in certain economic and social areas during the reform era, state intervention has not been totally withdrawn.

Essentially, the government has shifted from direct to indirect surveillance (Tang and Chung, 2000). Although privatisation has been promoted in most state-owned enterprises (SOEs) in the industrial sector, ownership is retained in joint ventures and private companies with intimate ties to the state. For instance, the Ministry of Chemical Industry owns and operates a number of chemical plants which have been granted import and export licences, and assigned quotas (Mastel, 1997, pp.68-69). Through this mechanism, the import and export of particular chemical products have been manipulated by certain plants with close state connections. Also, distribution channels have been freed but the state's control still persists through joint-venture arrangements, and it continues to play a crucial role in the allocation of commodities, particularly producer goods. Moreover, state authorities continue to intervene in the economy through the control of the banking system, taxation and loans. A double track system (*shuan gui zhi*), which refers to the juxtaposition of state control and market forces and planned and unplanned channels, has emerged. In the agricultural sector, all rural markets are operated and organised by the Industry and Commerce Management Bureau (ICMB). Peasants are allowed to sell most of their products in the market at prices set by supply and demand. However, special products, such as cotton, tobacco, silkworm cocoons, and pine resin, cannot be sold through rural markets.

Although reform has not dismantled the state's control over the economy, a 'localisation' process has occurred in which the government's command structure has been decentralised to local government (Chung and Tang, 1997). Reform measures, such as the delegation of decision-making power and fiscal responsibility systems, have redefined relationships between: (a) central and local governments; (b) between local governments at all levels; and (c) between different local political and economic institutions. Local government is no longer the state's sole redistribution unit which involves either collecting commodities from the local areas or allocating them from a central agency. Clearly, the reforms have re-interpreted local government's tax revenues, the ability to retain profits, property rights and authority to allocate resources. Consequently, a considerable number of plants and properties have been transferred to local control. Local government investment and financial management powers have been reinforced through the fiscal responsibility system, local banks and other credit unions are now under their control. Moreover, local government administrative territories and economic hinterlands have been expanded by the implementation of city-leading-counties (*shi guan xian*) (Chung and Tang, 1996). These arrangements have not only reinforced local government administrative powers, but also stimulated them to expand into activities previously constrained by capital and material shortages during the pre-reform period. While these reforms have prompted local governments to improve their revenues, they have also increased their desire to intervene in local economic affairs.

Intervention has been spurred by both economic and social factors. Economically, decollectivisation has stripped local government revenues and reduced the salaries of cadres provided by the People's Communes. The budgets of local governments have been further constrained by reforms, forcing them to seek other financial sources to meet budget needs imposed by higher levels of government. As any profit and revenue over the assigned amount is retained by local government, maximising the returns of district enterprises offers an opportunity to boost income. Socially, local government also depends on these businesses to contribute towards community employment, social welfare and housing. All of these have been sensitive factors during the reform era. Social and political unrest will result if these objectives are not attained.

Although the autonomy of firms has been increased by the decentralisation of economic decision-making power, their subordinate relationship with government remains unchanged. A mechanism for government intervention still persists. As enterprises depend on the state for markets, supply of resources, investment and subsidies, they do not necessarily resist intervention by local government. A 'dual dependence' association between the two has developed (Walder, 1995; Jiang and Hall, 1996). By intervening in the affairs of local enterprises and other aspects of the economy, local government has developed a dual role as administrator and entrepreneur (Oi, 1992, 1995, 1999b; Nee, 1992). The term 'local corporatism' has been coined to describe this development (Oi, 1992).

Several studies have discussed local corporatism. Nee and Su's (1996) study of Xiamen city – one of the 14 coastal open cities – details this concept. In the 1990s, Xiamen's government and 3 local, collectively-owned firms established a joint investment company to promote exports. The government not only contributed 60 per cent of the company's capital investment, but also appointed the board of management. According to Nee and Su (1996, p.120), the Xiamen city government's share in these 'window companies' has ranged from 10 to 100 per cent. Moreover, the Xiamen city government insists that local banks grant loans to these businesses and acts as a guarantor to raise capital in both domestic and international capital markets.

The overlapping of administrative and economic roles is more obvious at lower levels of governmental administration – township and village. Generally, cadres in townships and villages retain the role they played in the People's Communes, managing collectively owned enterprises and distributing resources to villagers (Watson, 1992, pp.182-188). Consequently, it is not surprising to see the local party secretary or bureau head as the general manager of township and village enterprises (TVEs). Other scholars, notably Blecher and Shue (1996), Qian and

Stiglitz (1996), Brown (1998) and Marton (2000), also provide evidence of this mutual dependence. For instance, the involvement of local government in major entrepreneurial decisions and the mobilisation of resources and capital in shaping enterprise development in various parts of China.

Amalgamation of the local government's administrative and economic roles has significant consequences for China's regional economic development. Tension between resource-exporting provinces and manufacturing provinces has been intensified. During the pre-reform period, a spatial division of labour between interior and coastal provinces was established (Yang, 1997). These interior provinces, which are rich in natural resources, specialised in exporting raw materials whereas coastal provinces concentrated on manufacturing. The allocation and distribution of raw materials and manufacturing goods between regions were strictly manipulated by the state's distribution and material apparatus through 'planned prices'.[2] Generally, the pricing system, however, has under-valued raw materials and over-valued manufactured and consumer goods.[3] In other words, provinces exporting raw materials were underpaid for these resources and overcharged for imported manufacturing goods. Under this unequal exchange, huge amounts of coal, crude oil and other materials were extracted from Xinjiang, Shanxi, Yunnan and other provinces. Conversely, Shanghai – the largest manufacturing centre – and other industrial cities benefited from this practice.

Decentralisation of decision-making power has increased the role of local government in local affairs. Given the unequal exchange and financial pressures stemming from fiscal reforms, raw material exporting localities have intensified their efforts to establish local resource-processing industries. This has given rise to the *zichan, ziyong, zixiao* ('local production, local use and local sale') in the wool producing areas of Gansu, Qinghai, Xinjiang and Inner Mongolia to stop the leakage of assets (Yang, 1997, p.69). Similar situations have been observed in other areas such as Shanxi, Heilongjiang, Xinjiang and Shandong. These actions have affected factories in net importing areas which are finding it more difficult to obtain the required raw materials and markets. As competition has intensified, tensions have increased between local governments, some of which have not hesitated to protect their interests by renewing administrative protection measures prevalent during the pre-reform period, and inherent in the administrative system of the

---

[2] The 'planned price' was set by the state under the system of 'unified purchase and unified sale', which was implemented in 1953.

[3] The pricing system's objective was to increase state accumulation. By reducing the production cost of the industrial sector, net profits and revenue to the state were increased. Consequently, interior resources were pumped to the coastal area through this system.

reform era. These have included taxes, embargoes, roadblocks and even military action.

The outcome of government intervention is market fragmentation. This phenomenon has stemmed from 'commodity wars' in different parts of China over raw materials such as wool, tea, tobacco leaves, silk and cotton. Although these activities lessened during the 1990s, other administrative measures have been used to protect markets. A detailed case study of this is given in Chapter 6. The prime reason for market fragmentation is the relaxation of central control. As the governmental command structure has been decentralised, the administrative and entrepreneurial roles of local government have been blended. Local governments have not hesitated to use administrative measures to protect their interests when these are jeopardised.

Regional conflicts have not only occurred between material export and manufacturing areas but also between different administrative units. Driven by the need to attain high profits and revenues, local governments launch projects which promise high returns within their borders. Assembly lines for consumer goods have been imported extensively. In the 1980s, duplication was witnessed in tape recorder, electric fan, colour television, washing machine and refrigerator manufacturing in different cities and counties. This situation continued into the 1990s with most investment being replicated in automobile assembly lines, petrochemical industries, video-CD player manufacturing and 'development zones' (*kaifa qu*). Again and again, administrative boundaries have been used effectively to protect the local economy – particularly when there are shortages in the supply of raw material and when the central government implements tight measures to slow down the over-heated economy and inflation.

Consequently, China's regional economy has taken on a cellular structure. Each 'cell' comprises the administrative zone of local governments such as a province, city, county and town. Within each 'cell' there is a small empire of similar industrial and economic structures. Instead of integration and cooperation, these administrative zones make their own independent policy, compete vigorously and are hostile to each other. A special 'administrative-zone economy' has thus been created (Liu and Shu, 1996; Liu, 2002). By 2000, there were more than 600 cities and over 2,000 counties in China. This administrative-zone economy implies that China's market is divided into thousands of small pieces, despite the state's attempt to co-ordinate activities as if 'the country is a single chessboard' (*quan guo yi pan qi*).

The administrative-zone economy cannot be seen merely as the spatial expression of local corporatism. It is also the product of China's 'half-baked' reforms – a great leap in economic matters but only a tiny step in political aspects. The great leap in economic reform has contributed to remarkable changes in the rural, industrial and financial sectors, and the noted formation of local corporatism. The tiny step in political reform, however, has retained Mao's administrative hierarchy. Consequently, the governing mechanism inherited from the reform era still ensures the operation of the administrative-zone economy.

The administrative hierarchy is probably the most significant measure established in Mao's development model, which merged the country's political and economic hierarchies. Accordingly, there was only one ranking in China – the political-cum-economic hierarchy regarded by Skinner (1964) as an 'artificial' system. Political and administrative functions were over-emphasised in Mao's hierarchy while economic concerns were undermined. Under Mao's system, all central places, such as cities and towns, were regarded as administrative bodies rather than economic centres. 'Cities' and 'towns' thus became designated administrative organisations and levels of government in the hierarchy. All central and local governmental authorities, such as police, procuratorial, legal and judicial departments, and People's Congress, were situated there, although there were no walls and a moat surrounding cities like those in Imperial China.[4]

The above situation also explains why the term 'city' has been so confusing in China. First, the city is not only a central place in the economy but also a level of government. Secondly, there are different levels of city government such as municipality, prefecture-level city and county-level city (Figure 3.1). As administrative activities are paramount they determine the city's economic functions and powers; higher ranks in the hierarchy possess the greatest power. For example, the major difference between prefecture-level and county-level cities has been that the former enjoy the authority of promulgating decrees but the latter do not (Pu and Zhu, 1993). Moreover, prefecture-level cities are entitled to establish urban districts (*shi qu*) within their jurisdiction, but not county-level cities. The population within such districts is regarded as an 'urban population', which enjoys better social welfare and funding from the central government than its rural counterpart, despite the malfunctioning of the household registration system during the reform period.

The political-cum-economic system reached its extreme expression when the People's Communes were established in rural China. Although most proved to be disasters and were finally abolished, the practices of administrative control have persisted in urban areas. During the 1960s and 1970s, there were between 170 and 180 cities in China, demonstrating that the number of urban administrative units

---

[4] Please refer to Fei (1953, pp.95-104) for a comparison between political and economic centres in Imperial China.

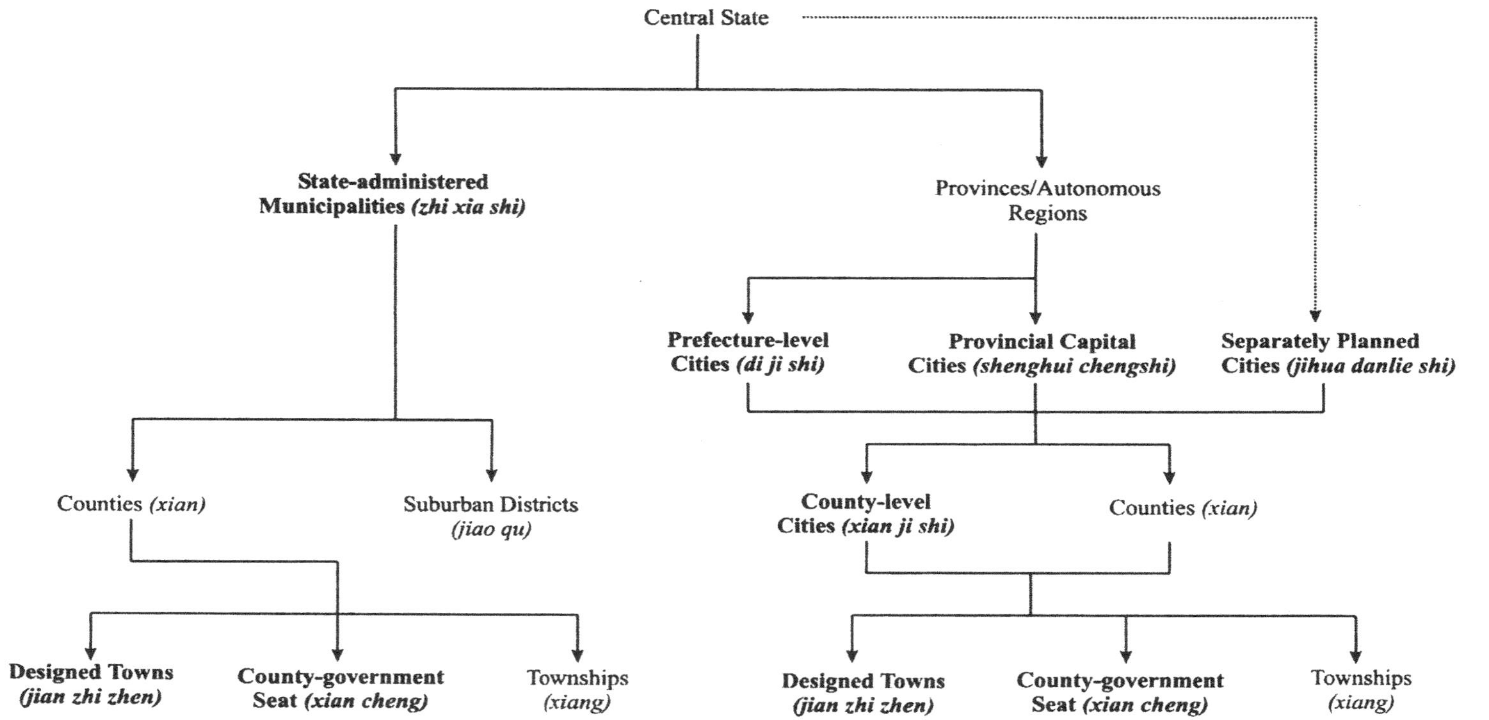

Designated cities and towns are in bold

Figure 3.1 China's administrative hierarchy since 1982

were maintained at about the same level before economic reform commenced in 1979.

Reform has done very little to separate the political-cum-economic hierarchy. Among the 31 provinces and autonomous regions in China, only Hohhot city – the provincial capital (political centre) of Inner Mongolia – is not the province's leading city in terms of economic indicators. The other 29 provincial capital cities still exhibit the strong integration of political and economic functions. A significant number of economic central places have been assigned the administrative status of city or town to accommodate the rapid development of new urban areas.[5] For example, when the status of 'city' is granted to a county, the amount of tax revenues retained for local construction is expanded from 5 per cent to 7 per cent (Liu, 1991). Also, local government is endowed with greater power in allocating local goods and materials, approving land uses and adjusting prices. Moreover, investment from the central government is increased and grants and loans offered. Thus, economic benefits are closely related to administrative status.

During the 1980s and 1990s, there has been a rapid increase in the number of cities and designated towns. In 1978, there were only 193 cities; by the end of 2000, stimulated by the reform era's economic liberalisation, the number had soared to 663 (SSB, 2001, p.3). The total of designated towns has also increased from 2,851 in 1979 to 7,511 in 1985, and 19,692 in 2000 (SSB-RSST, 2002, p.31). Distinctive status, such as special economic zones and separately planned cities (*jihua danlie chengshi*), were also used to nominate economic central places which were neither primacy cities in a political sense nor provincial capitals. These developments have given special status areas economic decision-making powers equivalent to those of provincial-level units. Designation as a city or town means an economic central place has been 'institutionalised' and becomes part of the administration system. The institutional process has not only allotted cities and towns political functions but also endowed them with a certain level of economic discretion. In the era of economic reform, these powers have awarded cities and towns certain privileges, including the ability to seek foreign investment. These events explain why economic central places have to be promoted administratively before they can undertake major reform initiatives. Concomitantly, if the economy of a rural county expands sufficiently, the rural county is promoted to the rank of 'city'.

The political and economic systems are regarded as a single hierarchy with overlapping spheres of interest under the administrative-zone economy. Over time the administrative system takes precedence over the economic one. The zone's

---

[5] Based on this observation, Chung and Tang (1997) have argued that China's urbanisation is a "totalisation process", in which more and more space is absorbed into the administrative system and subject to state control.

boundaries also encompass its economic hinterland and, within this jurisdiction, local government has absolute power. This political-cum-economic structure has several implications for China's regional economic development, including: (a) the alignment of administrative with economic central places; (b) the dominance of local government within the designated administrative zone; (c) the paucity of interaction between administrative zones due to competition and protectionism; and (d) the plight of residents in border areas due to their peripheral location.

Thus the administrative-zone economy is a unique product of China's gradual reforms, marked by the decentralisation of power and 'half-baked reform'. Decentralisation has redefined the relationship between the state and local government, and a significant number of enterprises are now subordinate to local government. Consequently, the local government's financial power is reinforced, which, in turn, has increased local government's desire to intervene in the local economy. Further, high-level government intervention has also been retained because the hierarchical administrative system has remained unchanged during the reform era. Administrative-zone economy is also regarded as the product of rudimentary market forces. The juxtaposition of state power and market forces has enabled local government to use administrative measures to interfere in economic activities. Conflicts between state control and market forces will not cease until a mature market is developed. However, given China's gradual reforms, a relatively stable balance of state and market powers has been created. This situation, according to Liu and Shu (1996), has contributed to the stability of the administrative-zone economy.

Rural market development in the China context

Commercial activities in China have experienced considerable growth during the reform period. During the early 1980s, the number of market-places grew because restrictions on trading in agricultural products were abolished and non-state agencies and individuals were allowed to participate. A remarkable expansion in the number of market-places occurred between 1984 and 1989 with the abolition of unified purchases of agricultural sideline products and recurrent increases in state purchase prices (Tables 3.1 and 3.2). During this period, market turnover value recorded an average annual rise of 26.5 billion *yuan* (Cao, 1996). In prosperous coastal areas, the traditional periodic markets have been replaced by daily markets.

Table 3.1 The number of rural markets established since the reform era

Year	Total no. of markets	Urban	Rural
1979	38,993	2,226	36,767
1980	40,809	2,919	37,890
1981	43,013	3,298	39,715
1982	44,775	3,591	41,184
1983	48,003	4,488	43,515
1984	56,500	6,144	50,356
1985	61,337	8,013	53,324
1986	67,610	9,701	57,909
1987	69,683	10,908	58,775
1988	71,359	12,181	59,178
1989	72,130	13,111	59,019
1990	72,579	13,106	59,473
1991	74,675	13,891	60,784
1992	79,188	14,510	64,678
1993	83,001	16,450	66,551
1994	84,463	17,890	66,569
1995	82,892	19,892	63,000
1996	85,391	20,832	64,559
1997	87,105	22,352	64,753
1998	89,177	24,127	65,050
1999	88,576	24,983	63,593
2000	88,811	26,395	62,416

Source: SSB (1999, p.553), SSB-RSST (2002, p.207), SSB-GMSD (1999, p.233).

Table 3.2 Rural market turnover value, 1979-2000 (in billion *yuan*)

Year	Total turnover value	Urban	Rural
1979	18.3	1.2	17.1
1980	23.5	2.3	21.2
1981	28.7	3.4	25.3
1982	32.8	4.1	28.7
1983	37.9	5.1	32.8
1984	45.7	7.5	38.2
1985	63.2	12.0	51.2
1986	90.7	24.4	66.2
1987	115.8	34.7	81.1
1988	162.1	54.5	107.5
1989	197.4	72.4	125.0
1990	216.8	83.8	133.0
1991	262.2	108.0	154.3
1992	353.0	158.3	194.7
1993	534.3	256.2	278.1
1994	898.2	457.0	441.2
1995	1159.0	617.6	541.4
1996	1469.5	788.3	681.2
1997	1742.5	946.9	795.6
1998	1983.6	1104.3	879.3
1999	2170.8	1232.5	938.2
2000	2427.9	1380	1047.9

Source: SSB (1999, p.553), SSB-GMSD (1999, p.233), SSB-RSST (2002, p.207).

Strong growth in market and commercial activities continued in the 1990s. Between 1978 and 2000, the number of market-places had expanded threefold to 88,811, and with turnover value increased 193 times to 2427.9 billion *yuan* (SSB-RSST, 1999; 2002, p.207). Over 63 per cent of the country's total retail sales of consumer goods were derived from market turnover values. Among the 88,811 market-places in 2000, 74,604 were classified as 'comprehensive' providing daily necessities such as foodstuffs, clothes and household articles (DTM, 2001, p.373). Specialised industrial and agricultural product markets accounted for 4,727 and 7,439 respectively (DTM, 2001, p.373). Thus the market-places have replaced state distribution agencies and become major exchange channels. It appears that the rural market system has revived.

Despite the dramatic rise in the number of market-places and the value of turnover at the national level, the growth of rural markets is still far below the equivalent urban areas. Table 3.1 shows that the average annual increases in the number of urban and rural markets were 9 per cent and 1.7 per cent respectively, over the period 1985-2000. During the same period, the value of market turnover in urban areas increased at an average annual rate of 46 per cent. Conversely, only 26 per cent of annual growth were recorded in rural areas (Table 3.2). Given the bulk of the expansion in both numbers and turnover in the market-places was in the urban areas, a substantial development gap between the two is indicated. With rural areas accounting for about one-third of China's landscape and over 70 per cent of its population, rural markets are indeed underdeveloped, despite the liberalisation of circulation of goods, and local incentives.

A comparison of Skinner's estimate of the number of Chinese rural markets with current official figures provides further confirmation of the lack of development in these markets. Skinner (1965a, p.226) used a geometric model to show that in 1948 there would have been 65,000 'standard markets' in China; this figure was then adjusted downward by 10 per cent, to 58,555, after taking the modernisation of markets into account. The official figure for 2000 was 62,416; thus both the Skinner figures and the official figures are of the same order of magnitude. If both figures are relied upon, the rate of growth in the rural markets over the five decades (1948-2000) is less than 3 per cent per decade, or 0.6 per cent per annum. The increase in rural population, on the other hand, was 60 per cent between 1953-2000 – from 503 to 807 million – a growth rate of 1.25 per cent per annum (SSB, 2001, p.91). Since market growth depends upon population increase, there is clearly a great discrepancy between the respective growth figures.

The underdevelopment of rural markets can be explained by local government behaviour under the administrative-zone economy. Due to financial pressures induced by the fiscal responsibility system, it has concentrated its efforts on the industrial sector and property development because these generate larger profits and revenues. Generally, facilities that support agricultural production, such as water conservation, have been ignored and local government contributions to the agricultural sector slashed. In 2000, local government provided only 2 per cent of total expenditure towards agricultural production (SSB, 2001, p.266). This figure is lower than the central government's input to the agricultural sector, which was 8 per cent of total expenditure in 2000 (SSB-RSST, 2002, p.77). However, the real amount is even smaller than these figures because local governments divert some of their grants to other purposes.

As noted, decentralisation has not only given local governments more powers, but also has expanded their territories. While there is less central government involvement during the reform period, rural areas are subject to more direct and powerful controls from local governments. Rural resources have been extracted through various vertical and horizontal channels established during the reform era to generate profits in specific sectors (Chung, 1994). The 'white slips' (*bai tiao*) provide the best example. These papers of debt (IOUs) have been given to peasants by the local authorities instead of paying them cash for their grain; in many cases, peasants eventually receive nothing. Another example is the heavy burden of taxation and local levies. Between 1994 and 1997, the 'big three' burdens – agricultural taxes, assigned quotas and village allotment payments – accounted for about 8 per cent of a peasant's cash income (CASS-RDI and SSB-RSESG, 1998, p.15). Yet this is only the national average; in more remote areas or in counties renowned for their poor economic performance the proportion is even higher. Other instances have included the requisition of farmland for profit-making activities such as property development and the creation of industrial zones.

The slashing of agricultural inputs and leaking of rural resources have had significant impacts on rural market development. Lower agricultural inputs mean that there is insufficient capital to establish rural markets which explains why most of them are old with primitive and temporary facilities. The decrease in government inputs to the agricultural sector has affected not only the establishment of markets, but also the production and income of peasants. Since rural markets have a close and direct relationship with peasant production, any decline in productivity and income will reduce consumption. Eventually, rural commercial activities and market development suffer, in turn affecting the sale of agricultural surpluses by peasants. In the 1990s, many agricultural products experienced large-scale 'selling problems' throughout China. Indeed, recent Chinese evaluations of rural market development have pinpointed the importance of productivity and income, and their impact on the growth of rural markets (Chen, 1997; Gong, 1998; Han, 1998; Zheng and Zheng, 1998; Yan, 2002). These studies argue that the low consumption rate of peasants is the major reason for the underdevelopment of rural markets in China. As their real consumption expenditure increased by only 0.4 per cent in 1997, it is very difficult to expect rural markets to flourish in the countryside within a short period of time (Han, 1998, p.6). Increasing the income of peasants is regarded, therefore, as the state's top priority to promote a flourishing rural market.

Under the administrative-zone economy's political-cum-economic structure, the establishment of rural markets is determined primarily by administrative factors. Rural market-places under the present system must have market licences issued by the Industry and Commerce Management Bureau (ICMB), a state authority that manages all commercial activities, including rural market development. The licence system implies that any market-places and street peddling outside the system are regarded as illegal and suppressed by the Bureau. It also highlights that Skinner's (1964) 'natural' marketing system, which was formed solely by market forces, no longer exists. Under the licence system, there must be a bona fide licence holder or owner. According to Shi (1995), there are five types of market ownership. They are: (a) state; (b) departmental; (c) enterprise; (d) private; and (e) joint (by the department and other parties). The first two are generally regarded as state-owned, and include over 90 per cent of rural retail markets and a considerable number of wholesale and specialised markets. Enterprise and privately owned markets are usually found at the village level, particularly in coastal areas where both rural and township enterprises are flourishing. In the countryside, the majority of rural markets are owned by the local ICMB, meaning that not only is the Bureau the supervisor of rural markets, but also their owner and operator (Plate 3.1). Establishment of markets, selection of sites, requisition of land, construction of sheds and stalls, leasing of stalls, levying of fees, and management, are all manipulated by the Bureau. It is also known to lease the most desirable stalls to individuals with whom it has close relationships.

The intimate connection between the ICMB and rural markets suggests a Bureau must set up before market-places can be established. As the Bureau is an administrative unit found at different levels of government, such as provinces, cities, counties and towns, with various size and power, the attached outlets display a range of scales and functions. However, rural markets may also be established in townships (*xiang*) which are not recognised as a designated government level within the administrative system. In this case, they are subordinated to the ICMB located in a designated town. This practice demonstrates that there can be no natural growth of market-places under China's political-cum-economic system.

Consequently, the market hierarchy reflects the organisational structure of the ICMB and levels of local government. In general, small markets – or standard markets in Skinner's (1964) terms – are established in designated towns, although this is not always the case. Many standard markets are periodic markets which perform only simple exchange and retailing functions for peasants. A town's population size is not a major criterion; irrespective of whether it has a population of 5,000 or 30,000 it is served by one rural market. The town's administrative territory is also the market's hinterland. Where designated towns are town-government seats, there could be two rural markets. Usually, the second one is an intermediate, large-scale daily market with a greater variety of commodities and services, particularly high-order goods. It would also provide basic wholesale functions for standard markets. Central markets are established in provincial capitals, usually a prefecture-level city. In some cities, wholesale markets are also present. Such establishments demonstrate that the rural market hierarchy is still closely aligned to the administrative one.

At the local level, for administrative purposes, rural markets are always placed in the town-government seat which houses different government bureaux and offices. Precisely, market-places are always adjacent to the ICMB office, the authority which controls their development (Plate 3.2). This position enables Bureau officials to patrol rural markets during market days and collect fees from peasants. Although the town-government seat has a concentration of non-agricultural population, it is not an optimum situation for peasants as often they have to travel long distances to visit the market-place. This problem is compounded in the hilly or mountainous areas where transport facilities are primitive and roads poorly maintained. Moreover, the town-government seat is not always situated in an ideal spot, some are far away from accessible main roads and others are located along defunct waterways or routes that are usually bypassed.

The relocation of the town-government seat is always accompanied by the re-siting of the rural market, although relocation to accommodate market forces is rare. In this sense, the demise of a rural market in China is not solely due to transport factors, as argued by Skinner. The location of the town-government seat also constrains market expansion because available land there can be very limited and is generally occupied by different authorities. This means that market activities can only expand through street peddling, which is scattered, and economies of scale cannot be obtained. In Guangdong Province, a furniture market at Shunde extends along a road for over 5 km and while a roadside location is advantageous, it also creates traffic problems.

Market days and market size are not determined by administrative factors. Generally, market days follow the traditional pattern authorised by the ICMB. However, the lunar calendar has been abandoned despite its widespread use among peasants. Like traditional periodic markets, neighbouring outlets have different schedules, allowing peasants to visit different rural markets without any time conflict. Although size is not determined by administrative factors, it is always controlled by local government. As noted, lack of capital input by local government has contributed to the underdevelopment of market-places. Consequently, many venues are small in scale and their facilities old and inadequate (Plates 3.3 and 3.4). However, there are local governments which have established extravagant market-places to demonstrate their economic performance and financial achievements. For example, in Guangdong's coastal areas there are huge multi-storied rural markets. Whether a market is shabby or extravagant is a good indicator of the mismatch between its scale and the size of its economic and population capacity. Sometimes, these situations have created a 'market without a market-place' (*you shi wu chang*) and a 'market-place without a market' (*you chang wu shi*) (Cao, 1996). These situations have definitely stemmed from the intervention of local government.

Plate 3.1 The main entrance of a rural market – all government markets have this distinctive gateway

Plate 3.2 A typical rural market – ICMB office (the building) is attached to the market-place (ground floor)

Plate 3.3 Street markets still exist in some remote areas

Plate 3.4 A scene at a rural market – many standard markets are old with primitive facilities

Wholesale markets are another topic that characterise China's market development during the reform era. Although Skinner does not address wholesale markets in his studies, it is a key feature of the process of market modernisation. In 1986, there were only 892 agricultural wholesale markets and turnover value was 2.8 billion *yuan* (Li, 2002). By 2000, their number increased to 4,532, with a turnover value of 335.1 billion *yuan* (DTM, 2001, p.379). A breakdown of their types and turnover value is shown in Figures 3.2 and 3.3. The first wholesale market in China was a vegetable market established during the mid-1980s in Shouguang county, Shandong's major production area. By the early 1990s, various wholesale markets had been set up in different cities such as Beijing, Shanghai, Guangzhou, Chengdu, Zhengzhou and Tianjin. Table 3.3 lists these wholesale markets. Like retail markets, these emerging wholesale markets are also controlled by the state through a licensing system. According to Watson (1996), wholesale markets have been established in three ways: (a) by the ICMB; (b) by the state's commercial agencies; and (c) by independent organisations and managed by the Bureau. Although no detailed statistics are available, a study conducted by the Agricultural Ministry (1995) shows that a significant number of wholesale markets were jointly established by local governments and the ICMB. According to Xu (1996), local governments have regarded their 'assistance' in establishing wholesale markets as a matter of public policy. Local governments and other local authorities have intervened in the operation of these markets by providing both capital and land.

Table 3.3 Major wholesale markets for agricultural products in China

Location	Trade items	Establishment agency	Major trading unit	Set up date
Beijing	Grain	Grain Bureau of Beijing city	State-owned enterprises	1993
Shanghai	Pork	Shanghai city government	State-owned enterprises	1993
Changchun	Maize	Jilin provincial government	State-owned enterprises	1991
Chengdu	Pork	Sichuan provincial government	State-owned enterprises	1991
Zhengzhou	Wheat	Domestic Trade Department and Henan provincial government	State-owned enterprises	1990
Tianjin	Sugar	Domestic Trade Department and Tianjin city government	State-owned enterprises	1992

Source: Agricultural Ministry (1995).

Figure 3.2 Proportion of rural wholesale markets by type, 1997

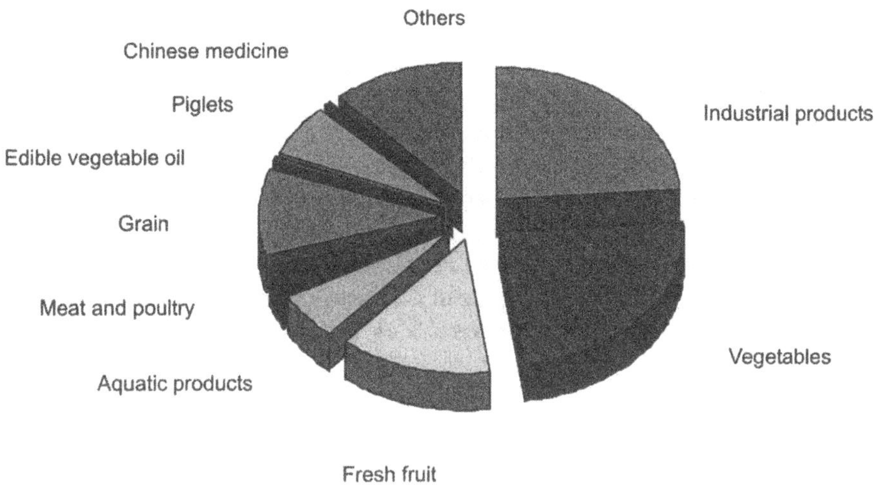

Figure 3.3 A breakdown of turnover value for rural wholesale markets, 1997

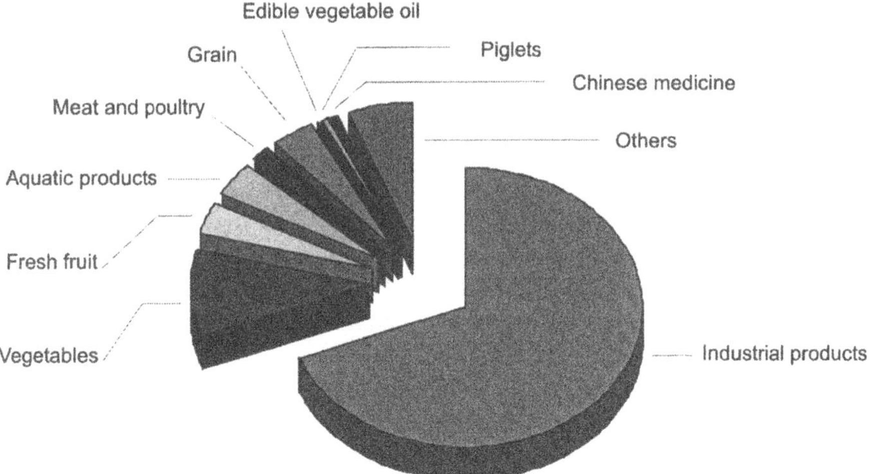

Wholesale markets controlled by government do not give free rein to economic forces. Usually, intervention occurs through price controls and subsidies. Zhengzhou's state-run wholesale grain market is a good illustration. First, price ranges for auction and negotiation – the two price-setting systems used in the wholesale markets – were determined by the local government (Xu, 1996). Then the local government compensated for any contract prices which differed from current spot prices by more than 6 to 10 per cent (Xu, 1996). This arrangement protected mid and long-term contracts, which accounted for one-third of those signed every year. Also, to protect local interests, direct administrative measures were introduced to control the demand and supply of commodities. This has severely interfered with the operation of market mechanisms and market integration. For instance, the persistence of departmental and regional trade barriers has fragmented sugar trading in northern China, despite a wholesale sugar market being established in Tianjin. As a result, a significant amount of the sugar trade occurs outside the wholesale market. Government intervention also contributes to the irrational geographical distribution of wholesale markets. For example, over 40 per cent of Guangdong Province's wholesale market is established in the Pearl River Delta – the southern part – while the vegetable and fruit production bases in the western and eastern parts have only a few of them (Hua, 2001).

As with the retail markets, local governments are not only the owners of wholesale markets but also participants. In many instances, private buyers are not strong enough to compete with government commercial agencies and corporations which have excellent transport and storage facilities and business networks. Under these circumstances, it is not surprising that one or two government corporations monopolise the markets in some places. In fact, a membership system has been adopted at state-run wholesale markets. Only members, usually state commercial corporations, are allowed to trade there. In this sense, it appears that government control is more severe in wholesale markets.

The markets formed under the political-cum-economic system do not conform to an economically rational pattern. According to official figures, comprehensive retail markets accounted for 84 per cent and wholesale markets 8.7 per cent of the 88,811 rural markets in 2000 (DTM, 2001, p.155). Many retail markets are merely standard markets providing only basic commodities exchange and retail functions for peasants and serve only small areas (i.e. a town) (Cao, 1996). Conversely, there are only a few retail markets with high-order functions. Given China's extraordinary scale and rapid economic growth during the reform era, the development of wholesale markets is still unsatisfactory (Li, 2002). Their pattern demonstrates the rudimentary and economically irrational nature of China's rural market system.

The political-cum-economic structure has redefined the functions of rural markets. They are now regarded as important tools to maintain social and political stability in the countryside. As the Chinese Communist Party's (CCP) success was based on a peasant revolution, the state clearly understands that an unstable rural society may have serious social and political consequences. This is particularly the case with the growth of surplus labour in rural areas following the implementation of the household responsibility system. Furthermore, the peasant's burden has been aggravated by various allotment payments and extra tax levies. Agricultural production costs are soaring and there are great fluctuations in the incomes of peasants. Rural markets are regarded as an effective tool to appease the unhappiness of peasants, providing them with working opportunities to increase their incomes and living standards. According to Shi's (1995, p.74) study, the 30 rural market-places in Baoji county during 1992 provided employment opportunities for 11,469 households in surrounding areas. Thus market-places have not only absorbed rural surplus labour but also alleviated serious underemployment in urban areas.

The Fifteenth Congress of the Party Central Committee in 1998 assigned another task to rural markets. Despite an 8 per cent growth rate being recorded, the regional economic and financial crisis between 1997 and 1998 severely damaged the country's exports and economic growth, particularly in rural areas. In order to guard against any future economic downturn and maintain long-term stability in rural areas, the state decided to accelerate rural market development so as to stimulate domestic demand and hence national economic growth. The Party Central Committee decided on four key issues to be targeted in the new millennium, they include: (a) stimulating the productive incentive of peasants; (b) increasing their consumption power; (c) stable market prices; and (d) stimulating rural market growth. Diplomatic reasons also exist. Since China has joined the WTO, a well-established rural market system will definitely increase the country's bargaining power on trade negotiation. Whether the above targets merely remain on paper is yet to be seen. Once again, rural markets and their development are regarded as tools to reach the state's political targets.

Rural marketing activities, in a broad sense, are still controlled by the state and different local authorities. Major crops, such as grain and cotton, are subject to price control, despite the relaxation of price controls for most crops. State procurement still exists. Cotton, for instance, is subject to unified purchase and monopoly sale by the state. Private trade is absolutely prohibited. Control of grain is not as strict as that of cotton. Nevertheless, peasants have to fulfil the state's procurement quotas before they can sell their surpluses at the rural markets. Any pre-procurement sale is regarded as illegal. Grain is acquired at a price set by the state, and although the set price increases, it is always lower than the market price. In 1995, the difference sometimes reached 0.6-0.8 *yuan* per *jin*, the largest variation during the 1990s. These price differences encourage peasants to sell their grain in rural markets. According to the *Nongmin Ribao* (Peasant's Daily) (10 June, 1998), peasants preferred to sell grain at rural markets not only because the price is higher, but also cash is paid immediately. Further, private dealers do not impose extra levies and grain deductions on peasants but local government does. This practice has jeopardised local government's procurement quotas which are assigned by a higher level of government. This is particularly the case when market prices are significantly higher than the state's procurement price. To fulfil the procurement quota, it is not surprising to see administrative measures, such as market closures, roadblocks and forced procurement, being adopted by local governments and state commercial units (Han and Yu, 1995, 1996; Zhou, 1994). Moreover, to safeguard their own interests, local governments have, on occasions, blocked the outflow of grain. Conversely, local governments have also reduced the procurement price illegally and used the difference between official and illegal prices to make huge profits. Such stories are not new in the rural areas. As reported by *Nongmin Ribao* (Peasant's Daily) (26 June, 1998), a local authority in Jiangling district of Jingzhou city, Hunan province, used an illegally low price of 0.9 *yuan* to purchase grain privately from local peasants during the acquisition of the summer crop in 1998. The grain was then resold to the state grain authority at the official price of 1.22 *yuan*. Consequently, each transaction of 100,000 *jin* earned the local authority a profit of 32,000 *yuan*. This is a typical example of local government intervention and misappropriation of state assets. It is not an exceptional case.

The same situation has prevailed in the procurement of special local commodities such as pine resin or silkworm cocoons. To fulfil quotas, administrative measures have been used to block the outflow of these commodities from production areas. Police forces have been involved, and peasants not only have had their products confiscated but also faced heavy penalties. However, unlike the 'commodity wars' of the 1980s, these blockades usually occur within local administrative zones such as counties. Usually, the conflict is between local government and the peasants. A detailed story of this activity is given later (see Chapter 6).

In sum, rural markets, both in the narrow and broad sense, are subject to intervention by local government and state authorities under the administrative-zone economy, despite the diminishing of the state's control through economic reform. In the narrow sense, the establishment of market-places, their number, location, scale and daily operation is controlled by various administrative factors. Owing to the political-cum-economic administrative system, the current rural market arrangement does not operate as a distinct and independent hierarchy. This arrangement has been heavily eroded by the administrative hierarchy and can only be seen as its appendage. Consequently, the current rural market hierarchy can only be justified in administrative and not in economic terms. This scenario is very different from the pre-1949 period. Although administrative and economic spheres sometimes overlapped, as described by Skinner, the political and economic hierarchies were basically two separate systems. In the broad sense, local government intervenes in market activities through price controls and direct administrative measures. Accordingly, economic forces have not been developed as smoothly as some economists have predicted. Again, these scenarios did not exist in the pre-1949 traditional markets described by Skinner.

Traditional rural markets appear to have been revived during the reform era. Consequently, one has witnessed: the withdrawal of state control in the distribution and trading spheres; the restoration of periodic markets and traditional market days; the remarkable increase in the number of market-places; a growth in market turnover value; and the emergence of wholesale markets. Moreover, daily markets have emerged in some prosperous regions, and long-distance marketing also occurs. Market-places have replaced the state distribution channels and become the major place for commodity exchanges.

This is not the end of the story. Abstract research, which employs oversimplified rules to study China's rural market development, has not gone beyond the above set of description. Consequently, improbable conclusions have been drawn. Thanks to the debate on the contextual approach and the importance of space, there is now a basis for concrete research on social and economic processes. By re-establishing the interaction between changes in rural markets and China's reform process, scholars have discovered that development is falling far behind its urban counterparts. A substantial gap between urban and rural areas has appeared therefore. Furthermore, the establishment of rural markets, their number, location and daily operations are closely controlled by the political-cum-economic administrative system and local government behaviour. Under these circumstances, the market hierarchy is also determined by the administrative hierarchy. In comparison, Skinner's (1964) praise for 'natural' economic centres in the per-1949 period shows that government interference in market activities was relatively weak then. While the current marketing system is closely controlled by government, the 'natural' economic system has not been restored.

The reconstruction of the relationships between China's market development and its political and economic contexts also demonstrates that the current market system has emerged from a unique set of spatial contingent conditions and causal effects. These contingent conditions and causal effects have been constructed in the reform era, when relationships between the state and local governments, and between different levels of local government have been redefined and decision making powers decentralised. This situation, again, is very different from the pre-1949 period, when: the country was still under the control of the Republican government; the national economy suffered from civil war; the state's capacity was weak; and the rural economy was primitive and closed. The traditional market system, as described by Skinner, was formed within this context. While some aspects of the current marketing system are similar to the traditional one, they are generated by different factors and operate within a different framework. In this sense, the traditional marketing system has not been revived.

Chapter 4

Local Market Pattern and Hierarchy: Case Studies of Deqing and Dongguan

Rural market development is both a function of a place's location and its social and economic progress. Given the variations in socio-economic development, the rural market system differs from place to place. This is particularly the case with China, which is characterised by its extraordinary scale and local diversity. Yet Skinner's mid-1960s thesis on rural market structure is problematic in this regard. There is no doubt that Skinner's two models of market structure have broadened the geographical horizons of scholars by extending central place theory beyond the homogeneous plain. However, his understanding of spatial variations remains centred on topographical differences – a narrow perspective which has weakened the explanatory power of his models. Physical landscape and transport are not the only elements that create spatial variations. Moreover, in an attempt to establish a grand theory that can "apply anywhere in the world", Skinner (1965a, p.17) cuts off the connections between rural market development and the changing dynamics of political and socio-economic contexts over time. Accordingly, his theory fails to anticipate new factors, such as foreign investment and rural industrialisation, which have had significant impact on China's regional development during the era of economic reform. This is demonstrated in the following discussion.

Deqing and Dongguan: A semi-closed economy vs. an open economy

Skinner's (1965a) two models of market structure and evolution in plains and mountainous areas suggest a simple correlation between three factors: topography, transport-productivity and market pattern. He argues that unfavourable landscapes, such as hilly and mountainous areas, generate poor transport and low agricultural productivity and hence create a scattered distribution of rural markets. Conversely, in the plains, better transport and higher agricultural yields lead to a much tighter market structure. However, the participation of the state, local governments and foreign investors in China's economy during the reform period suggests that a more complex correlation between these components and market development has been formed. In other words, variations in market structure and pattern are created by complex relations between these elements rather than by Skinner's simple equations.

Recent debates on the abolition of special economic zones suggest preferential policies are the major factor creating regional disparities during the reform era (Hu, 1999). The reform philosophy of 'let a small number of people get rich first' (*rang bu fen ren xian fu qi lai*) is indicative of the state's strong desire to regulate economic development via market forces. However, the central government has restricted preferential policies for foreign investment to a small range of cities and areas – a strategy which has created a marked gap in the level of development between Deqing and Dongguan in particular, and between coastal and inland regions in general. Although some neighbouring counties have benefited, Deqing itself has been largely excluded from foreign investment because it is located outside the open economic zone. Also, it is not a recipient of the special strategies planned by the provincial government in 1996 because it is outside Guangdong's Eastern and Western Wings' Region.[1] Conversely, Dongguan city is a core member of the Pearl River Delta Open Economic Zone and can capitalise on the range of programmes designed to attract foreign investment. This preferential policy instigated by the state is a major reason for the variations in market development.

Skinner has broadened scholarly understanding of external factors – notably transport routes and market modernisation. Yet he does not anticipate the significant impact of foreign investment on China's market development. Although foreign investment is not a key factor in establishing market-places, it affects their subsequent evolution by restructuring the industrial and agricultural sectors of specific places (Eng, 1997; Lau, 1998). Given the close relationship between agricultural production and rural markets, any significant changes in the former inevitably affect the latter. Industrialisation also influences market intensification by increasing the level of specialisation between industrial and agricultural sectors. Moreover, a successful transformation from an agricultural to an industrial economy includes an increase in income and density of demand – a change which also impinges on market development because individuals are major actors in its disposition.

---

[1] For a detailed report and documentation on the region, refer to Guangdong Province Planning Committee (eds) (1997) *Guangdong Sheng Dongxi Lianye Quyuguaihui Yinjiu (Study on the Planning of the Eastern and Western Ring Regions of Guangdong Province)*. Guangdong: Jingji Chubanshe.

Unlike Dongguan and other Pearl River Delta counties, which attract large amounts of foreign investment, Deqing has been shunned by foreign investors. In 2000, the total amount of foreign capital attracted was USD $13.8 million, ranking it sixth among Zhaoqing's six county-level units (ZQSB, 2002, p.284). In 1996, the county had 20 registered joint ventures employing 1,743 people which accounted for 7.3 per cent of its employment. In 2000, only two new foreign investment agreements were signed, fewer than anywhere else in Zhaoqing city. Moreover, a net outflow of population was recorded. In both 2000 and 2001, more than 7,500 people changed their household registrations to other places (DQSB, 2002, pp.15, 234). In addition, many people have temporarily left Deqing to become migrant workers in the Delta area. However, no detailed statistics are available on their precise numbers.

Preferential policies and central location have made Dongguan one of China's most attractive cities for foreign investment. In 2000, the total amount of foreign capital invested was USD $1.8 billion (GDSB, 2001, p.222). This ranked the city third out of the province's 21 prefecture-level cities. During 2000, 1,276 foreign investment contracts had been signed and 13,825 joint-venture enterprises had been established in the city (DGSB, 2001, p.108), including multinational companies such as Philips, NEC, Nokia and Nestlé. These firms have absorbed local surplus labour and attracted migrant workers from other areas of China. In 2000, over 60 per cent of Dongguan's 4.1 million population were migrant workers (DGSB, 2001, p.219).

Foreign investment has greatly bolstered Dongguan's industrial base in both urban and rural areas. Again, the influence of industrialisation on spatial variations in rural market patterns and development is another factor which Skinner's models do not anticipate. Between 1978 and 2000, Dongguan's gross value of industrial output (GVIO) increased from 0.6 billion to 49.2 billion *yuan* (DGSB, 2001, p.220). By 2000, industrial output accounted for 55 per cent of the city's GDP. Three quarters of Dongguan's 16,975 industrial plants were foreign-supported processing enterprises.[2] In 2000, 73 per cent of Dongguan's GVIO was contributed by these businesses (DGSB, 2001, p.314). A significant number of these are township and village enterprises (TVEs) which contributed over 70 per cent of Dongguan's total GVIO. State-owned enterprises, however, contributed less than 1

---

[2] Chinese materials refer to these industries as *sanzi* and *sanlai yibu* enterprises. *Sanzi* refers to three forms of foreign investment. They are: (a) equity joint venture; (b) contractual joint venture; and (c) wholly foreign investment. *Sanlai yibu* refers to four forms of production. They are: (a) processing imported materials; (b) assembling imported parts; (c) manufacturing according to imported designs; and (d) compensation trade. Both *sanzi* and *sanlai yibu* are the most common forms of investment in Dongguan and in the Pearl River Delta.

per cent. Thus, TVEs have replaced collective-owned firms and have become Dongguan's major source of taxes and revenue, leading to significant consequences for the rural areas including the spatial pattern of markets.

Conversely, the lack of foreign investment input in Deqing has resulted in stunted industrial development during the reform era. In 2000, industrial output accounted for 40 per cent of the county's GDP. Between 1949 and 2000, the total number of factories increased from 49 to 138. Of these, 110 were collectively owned and 12 owned by individuals (DQSB, 2002, p.355). Inevitably, these figures are overestimates because they refer to registrations not actual plants. Indeed, many state-owned factories have been downsized or have ceased operations. Those still functioning include several large, collectively owned installations – a silk reeling mill, wineries, pine-resin processing units and cement plants. Deqing's silk reeling mill, for instance, was one of the county's biggest enterprises employing over 2,000 people. Now less than 200 workers remain and its contribution to the county's tax revenue has dropped markedly. According to official figures, the number of state-owned and collectively owned plants accounted for 80 per cent of Deqing's total and their outputs made up 70 per cent of the county's GVIO. In contrast, although 51 per cent of the factories in the county were at the village level, their GVIO was only 15 per cent (DQSB, 2002, pp.248, 359). As post-reform industrialisation has generated few advantages for Deqing's rural areas, little change is observed in its market structure.

However, rapid industrialisation has accelerated Dongguan's market modernisation by restructuring its agricultural sector, an effect not anticipated by Skinner's (1965a) models which focused on agricultural production and marketing patterns in a pre-modern period. Since local industries have absorbed rural surplus labour and radically changed the living standards of peasants, the traditional dependence on farmland for one's living has been eroded. Agricultural production has switched from self-sufficiency to specialisation on cash cropping. Considerable areas of low-yield farmland have been converted to pasture. Also, paddy fields have been shifted into crops with high economic returns such as sugar cane, bananas and vegetables. Fruit, vegetables, aquatic products and poultry have been the most popular products. During the late 1980s, over 2,000 of these new farms were developed, some geared to overseas markets such as Japan and Hong Kong. Generally, they are large in scale, (i.e. over 1,000 *mu*), hire many workers, and use scientific methods such as new and enhanced species, soil-less cultivation and biological pest control. For instance, in Dongguan's Dalingshan town, a 3,000 *mu* orchard was established during the mid-1980s to specialise on fruit such as lychee, longan, oranges, mandarins and tangerines. Included in the orchard are a nursery, a fishpond and a pig farm. By the end of 2000, there were 7,510 of these specialised large-scale farms/households, and 235 farming enterprises, in Dongguan city, cultivating a total area of 184,542 *mu*, about 30 per cent of the city's arable land (DGSB, 2001, p.286).

This high level of agricultural specialisation has led to flourishing marketing activities within Dongguan city. Fruit, aquatic products and poultry are destined not only for the local market, but are also despatched from there to both national and international markets including Hong Kong, Macau and other Asian destinations. In Humen town, for instance, a 20,000 *mu* orchard and a 4,500 *mu* vegetable farm have contributed over USD $2.6 million in foreign exchange earnings to the local government. The town has also derived an annual income of USD $50,000 from its 3,000 *mu* aquatic farm. Since the mid-1990s, the commercialisation level of Dongguan's agricultural products has exceeded 80 per cent (DGSB, 1998, p.7). This suggests that self-sufficiency has no longer been the norm of peasant cultivation. As a result, retail as well as wholesale markets have been established – a modern market system is observed.

In comparison, Deqing's economy has yet to be restructured. Over 80 per cent of its population still depend on farming for their livelihood, the level of agricultural specialisation is low, and a small-scale peasant economy persists. Major farm crops include wet rice, sweet potato, potato, maize, soybean and peanuts that are primarily destined for the local market. Marketing activities are stagnant because exports from the county and internal transfers within China are rare. Consequently, Deqing's rural market growth has been constrained and its market structure has remained primitive with few departures from the traditional pattern.

Soil erosion and flooding are further crucial elements restricting Deqing's agricultural production. Given the close relationship between agricultural production and market growth, these constraints have also curbed market evolution. While Skinner's (1965a, p.196) thesis does cover these factors, little weight is given to them because they are regarded as exceptions: occurring only during periods of catastrophe in Chinese history. This generalisation applies to Dongguan but not to Deqing which has a long history of soil erosion and flooding. Soil erosion has become a serious problem in Deqing following the rapid increase of population, and massive deforestation for fuel and farmland. Not only has topsoil been washed away and soil fertility lost, but landslides and gullying have resulted. Farmland has been inundated and inland waterways clogged. Indeed, soil erosion has been regarded as the 'tiger that eats farmland'. According to a 1957 survey, Deqing's eroded area covered 378 km^2 – about 17 per cent of the county's territory (Deqing County Gazetteer Editorial Committee, 1996). About one-fifth of the county's hilly area suffered from soil erosion. Some 29,293 gullies were recorded within the county and 140,000 *mu* of farmland was endangered. Deforestation continued throughout the 1960s and 1970s as more land was opened up for farming and other purposes. An extra 36,000 *mu* of farmland was created but soil erosion worsened. Terracing was implemented to stop this but, in 1983, a second survey showed that the county still had lost 354.39 km^2 of land (Deqing County Gazetteer Editorial Committee, 1996). The removal of topsoil was particularly serious.

Reafforestation and other measures adopted since the 1980s have not completely stemmed erosion. Gullying and landslides have remained serious problems. These stem from the lifestyle of local peasants. 'Living on a mountain live off the mountain' (*qiao shan chi shan*) has been the Deqing peasant's philosophy of life for centuries. In an enclosed mountainous area, peasants depend on local resources for survival. Initially, jungles were cleared for housing and farmland; then trees were felled for fuel and timber sales; recently, hilly slopes have been excavated and soil and rocks taken for brick-making and construction. By concentrating on short-term benefits, peasants have overlooked the importance of the sustainable use of natural resources – this is a main cause of poverty.

Flooding is another factor constraining Deqing's agricultural production as the county is situated on the upper-middle course of the Xijiang River. The county's earliest flood was recorded in AD 996, when houses and farmland were submerged and peasants forced to leave (Deqing County Gazetteer Editorial Committee, 1996). Similar situations have been noted in historical records up to the present day. During the 1950s, large-scale water conservation projects were launched. Despite old dykes being consolidated, widened and heightened, and new concrete ones being built to protect the towns and villages along the Xijiang River, many of them – particularly those in rural areas – have not been strong enough to contain major floods. For example, major flooding occurred in 1994 when five towns along the Xijiang River – Yuecheng, Jiushi, Decheng, Huilong and Xinxu – were flooded and damage totalled 549 million *yuan*. In June 1998, another serious overflow again submerged the five towns and damaged the main road. External transport was suspended for eight days. Dykes in Huilong and Jiushi towns collapsed and caused severe damage to houses and farmland. Over 300 million *yuan* in losses was estimated. A detailed study of the impact of flooding on rural marketing is given in Chapter 5.

Soil erosion and flooding, coupled with declining arable land, have been the major environmental reasons for low agricultural productivity. Admittedly, paddy yields have never experienced rapid growth in Deqing county (Table 4.1). Between 1970 and 2000, the county's arable land decreased from 282,700 *mu* to 211,059 *mu* (DQSB, 1998, p.25; 2002, p.291). Although paddy yields have increased, they have still been marginal. According to the peasants, this is due to the use of chemical fertilisers which have had long-term harmful effects on soil fertility. During the reform era, most attention has been paid to the industrial sector and construction of infrastructure, with the agricultural sector being neglected. Government investment in irrigation and water conservation has been reduced. In addition to the soaring production costs, the incentive to grow grain has diminished. Yet the peasants cannot abandon cultivation because they have to fulfil the government's procurement quotas. Consequently, paddy yields are only sufficient for taxation and home-consumption. In 2000, Deqing county's paddy yield was the lowest among Zhaoqing city's six county-level units (ZQSB, 2000, p.180).

Table 4.1 Arable land and paddy yield in Deqing county

Year	Arable land (in thousand *mu*)	Paddy yield (in thousand ton)
1970	282.7	99.9
1975	273.5	115.0
1980	264.7	132.2
1985	238.5	112.6
1990	240.0	141.2
1995	210.2	127.7
1997	209.7	151.3
1999	210.1	143.3
2000	210.1	126.7
2001	210.1	120.1

Source: DQSB (2002, 1998, 1996).

While Dongguan's agricultural development demonstrates high levels of specialisation and export orientation, Deqing still remains relatively underdeveloped. Given the high level of self-sufficiency, rural markets persist with basic functions such as surplus exchange and retailing. No wholesale markets have been established as many products are consumed locally. During the 1970s, the county encouraged the cultivation of economic crops such as oranges, mandarins and tangerines. Since the 1990s, slopes have been opened for the growing of lychees, longan and mangoes. However, these advancements have neither successfully promoted cash cropping nor increased the level of specialisation. New varieties have not been as productive as those in the Delta area because of Deqing's adverse climate and soil. This attempt has also failed to augment the income of peasants. Like many places in China, peasants are attracted to fruit growing but invariably prices drop and their income shrinks. This situation has further restricted Deqing's desire to develop an advanced rural market system like Dongguan's.

Deqing's relatively poor industrial and agricultural advancement has constrained the growth of commercial activities, regarded as an important feature of market modernisation by Skinner (1965a, p.212). Apart from that, the level of commercialisation is crucial in causing spatial differences in market development between places. The higher the level of commercialisation, the more intense are interactions between people, locations and different sectors. These interactions provide the dynamics for shifting rural markets from the simple function of surplus exchange to modern trading.

Skinner suggests the expansion of commercial activities is one of the crucial factors for market modernisation, a perception which receives complete support from case studies of Deqing and Dongguan. Deqing's lower level of commercialisation means there are limited connections between the county and the outside world. Since many peasants depend on farming for their livelihood, they have little or no external associations because they do not need to travel to survive. Collectively owned plants are probably the only units with external alliances. Large collectively owned companies, such as the Forest Chemical Plant, account for over 90 per cent of the county's total exports which are controlled by the Native Product Import and Export Company, also a state-owned company. Although competition emerged when private firms were established during the reform period, they have always been too weak to challenge the Import and Export Company which has vast experience and better facilities and networks. Accordingly, the Import and Export Company has continued to dominate the market and most external connections. This explains why Deqing persists with a semi-closed situation and a simple market structure.

In comparison to Deqing, private companies in Dongguan have eroded the dominant position of state-owned enterprises in commercial activities. Before the economic reform, exports of agricultural side-line products, such as eggs, vegetables and poultry, were monopolised by the Dongguan Food Import and Export Company, the city's only state-owned food export enterprise. Like Deqing's Import and Export Company, this enterprise dominated Dongguan's external connections. Subsequently, the company's monopoly position has been challenged by the rise of private food trading firms during the reform era. Due to poor management and low efficiency, its market share has been greatly reduced like that of many state-owned enterprises. Now, the Dongguan Import and Export Company can only maintain its former market share in vegetable exports; movement of eggs, poultry, and some vegetables has been taken over by private food-export corporations. The company's decline in production is not rare and many state-owned enterprises in Dongguan have had similar experiences. Thus external connections are no longer the privilege of one or two big state-owned enterprises, as is still the case in Deqing county. With commercial activities booming in Dongguan, there has been an evolution of rural markets from a simple to a modern system.

Undoubtedly, transport has been one of the crucial factors in generating variations in market structure and modernisation. However, in China, it is the rapid post-reform development of commercial, industrial and marketing activities that is pushing transport advancement (World Bank, 1985, 1990, 1992). This causal relationship is different from Skinner's (1965a) thesis which suggests transport improvements inevitably result in market modernisation. The reason for this difference is China's lack of capital which has made transport a 'bottleneck' in local economic growth – a problem that was not resolved until foreign capital became available.

Dongguan's case is a good illustration. Since the mid 1980s, the government, in co-operation with Hong Kong developers, has built local highways to improve the transport system. However, the most notable event was the opening of the Guangzhou-Shenzhen-Zhuhai super-highway in 1993 which facilitated both the Pearl River Delta and Dongguan's development. As Dongguan is situated between the special economic zone of Shenzhen and the provincial capital of Guangzhou, it has benefited from all new highways and railways built by provincial government to connect these two places. In addition to the 4 major and 13 supportive roadways built in Dongguan during the past two decades, a relatively well-established road network has been put in place. In rural areas, many local roads have been paved. In 1997, the Humen river bridge was completed to link Shenzhen and Zhuhai – two special economic zones in Guangdong. Also, Humen port was upgraded and opened to foreign vessels. Improvements in the transport network have made long-distance trading possible as well as contributing to Dongguan's flourishing market activities and rapid modernisation during the reform era. This progress generally supports Skinner's argument connecting transport improvement with market modernisation.

In contrast, although Deqing had been linked to Zhaoqing and Guangzhou, the connections have been simple, unpaved roads. This poor transport system explains why a semi-closed economy with a simple market structure persisted in Deqing during the pre-reform period. Conscious of the Delta area's successful experience, Deqing county government has paid particular attention to roadway construction since the 1990s to boost economic activities. During the early 1990s, the Deqing government initially co-operated with a Hong Kong developer to improve the territory's roadway system, resulting in the Roadway Bureau becoming the county's most prestigious unit – particularly as its funds were bolstered by an injection of foreign investment. Indicative of this wealth is that the Bureau owns four, brand new, imported Japanese jeeps and one sedan. Conversely, the nearby Industry and Commerce Management Bureau (ICMB) has only a collection of locally assembled motorbikes and old bicycles.

In September 1995, Deqing's first road improvement project was completed. The 79 km Deqing section of the 278 km National Roadway No.321, which linked Guangzhou with Wuzhou – a major border city of Guangxi Province – was straightened and widened from 6 m to 16 m with 4 lanes, plus a 1 m footpath on both sides. This new road has shortened travelling time between Deqing and Zhaoqing city from 4 hours to 1.5 hours, and between Deqing and Guangzhou city from 8 hours to 3.5 hours. In 2000, another key project – the Xijiang bridge which links Deqing to its adjacent neighbour on the south shore of Xijiang River – was completed. This bridge is the second one along the river and opens up a new South-north path in the border area of Guangdong Province. However, despite the completion of these projects, many of Deqing's rural roads are still unpaved.

The Roadway 321 improvement and the Xijiang bridge have been the largest recent projects in Deqing. Local government hoped they would improve the county's market conditions and stimulate economic development by attracting foreign investment. Although these benefits have still to be realised, road tolls have contributed significantly to the county's tax base. Since 1998, toll payments in China have been legalised through the Roadway Law which allowed toll stations to be established with local and foreign capital. Two toll stations have been created on Roadway 321 in Deqing at Yuecheng and Decheng towns,[3] and one at the entrance of the Xijiang bridge. In 1997, toll payments form the former provided the local government tax revenue of 35 million *yuan*. Whether the boost in revenue will increase government's input into market development is unknown, therefore the impact of transport improvement on market modernisation has yet to be seen.

The lack of foreign investment, unsatisfactory development of industrial, agricultural and commercial activities, and relatively poor transport facilities account for the semi-closed condition of Deqing's economy, a condition which has both economic and political implications. Economically, Deqing's growth is slower than that of Dongguan; politically, government control persists in many aspects of its economic life because market liberalisation is less strong. Although the greater liberalisation of Dongguan's economy does not imply government regulation has been completely wiped out by market forces, it provides strong pressure to check and balance government intervention.

Deqing and Dongguan illustrate that Skinner's mountainous and plains models cannot fully explain the underlying causes which have contributed to a particular place's rural market pattern and evolution during the reform era. Topography and transport are not the only reasons for differences. During China's reform era, the prime causes distinguishing Deqing from Dongguan have been government policy, foreign investment input, level of industrialisation, and soil erosion and flooding. While Dongguan has been advantaged by foreign investment, Deqing, which does not enjoy any preferential policies, has been neglected. Foreign investment has helped restructure Dongguan's economy but not Deqing's industrial, agricultural and commercial base. While an open, export orientated and prosperous economy has developed in Dongguan, Deqing's economy has remained primitive, stagnant and in a semi-closed condition. The varying economic contexts of Deqing and Dongguan have thus created markedly different market structures and degrees of market intensification.

---

[3] Six toll stations exist on the 134 km roadway from Zhaoqing city to Wuzhou city, two of which are located in Deqing county.

Rural market hierarchy and development in Deqing and Dongguan

The contrasting levels of socio-economic, political and environmental development in Deqing and Dongguan are reflected in the relative growth rates of rural markets. In Deqing, improvement has been slow because of government regulation, its high level of self-sufficiency, semi-closed economy and relatively remote location. High levels of self-sufficiency imply low levels of specialisation and exchange of commodities. Although peasants do have surpluses to sell, these are small and irregular, and as they rely on their own farmland for survival there is little need for interaction with the outside world. In such circumstances, commercial activities and rural market development are stagnant, not to mention the additional burden of government regulation.

Deqing's stunted rural market development is reflected in the slow increase in the number of venues. The number of markets was small – only three – with the first record of rural markets being made in the sixteenth century. During the Republican period (1911-1949), this figure increased to 16 (Table 4.2). All of these had three market-days every ten days, for example, Huilong town on every third, sixth and ninth day of the Chinese lunar month (i.e. market days occurred on the 3rd, 6th, 9th, 13th, 16th, 19th, 23rd, 26th and 29th). Special schedules were set for the trading of specific commodities, for instance, pig markets occurred on every first, fourth and seventh day of the month, grain and coal markets on the second, fifth and eighth day, and silkworm cocoon markets on the third, sixth and ninth day. This situation is different from Taiwan – the only Chinese community in which Skinner's marketing framework had been applied – where periodic markets never existed (Crissman, 1976a, 1976b).

After the establishment of the People's Republic of China (PRC), the old market system and schedules were retained. However, as in most areas of China, rural markets in Deqing county experienced a major recession during the Great Leap Forward and the Cultural Revolution. Between 1954 and 1964, five markets closed and those which survived were strictly controlled by the state. During the 1980s, there was a gradual recovery and three new markets were established. According to the ICMB, Deqing's rural markets initially involved street peddling. Peddlers and customers were crowded into narrow streets or temporary sheds for buying and selling, market-places were dirty, stalls disorganised, and there were no rules and regulations to control and protect participants.

During the 1990s, two new markets were established and several old ones improved. Enhancements have included the construction of permanent concrete sheds and market complexes, and the implementation of regulations governing operations. Street peddling is only permitted in designated areas established by the ICMB. Within the market-place, a zoning system was introduced. At the end

Table 4.2 Number of markets created during different time periods in Deqing

Time period	No. of markets	Name of markets
Republican period (1911-1949)	16	Decheng, Guanxu, Maxu, Gaoliang, Mocun, Bozhi, Fengcun, Yongfeng, Yuecheng, Jiushi, Liucun, Huilong, Xiaoshui, Zacun, Shuixin and Shitou
People's Republic of China 1960	13	Decheng, Guanxu, Maxu, Gaoliang, Yongfeng, Mocun, Bozhi, Yuecheng, Fengcun, Jiushi, Liucun, Xiaoshui and Zacun
1975	11	Decheng, Guanxu, Maxu, Gaoliang, Yongfeng, Mocun, Bozhi, Yuecheng, Fengcun, Jiushi and Liucun
1985	14	Decheng Central, Decheng West, Decheng East, Guanxu, Maxu, Gaoliang, Mocun, Bozhi, Yuecheng, Yongfeng, Fengcun, Jiushi, Liucun and Shapang
1990	16	Decheng Central, Decheng West, Decheng East, Guanxu, Maxu, Gaoliang, Mocun, Bozhi, Yuecheng, Yongfeng, Fengcun, Jiushi, Liucun, Shapang, Datang and Wulong
1997	17	Decheng Central, Decheng West, Decheng East, Guanxu, Maxu, Gaoliang, Mocun, Bozhi, Yuecheng, Yongfeng, Fengcun, Jiushi, Liucun, Shapang, Datang, Wulong and Xinxu

Source: Fieldwork and Deqing County Gazetteer Editorial Committee (1996).

of 1997, there were 17 rural markets in the county (Figure 4.1).[4] Also, two farm cattle markets were established in Maxu and Yongfeng. In Huilong's town government seat a new market was due to open at the end of 1998 but this has still

[4] In Deqing, marketing data after 1997 is inconsistent because of administrative changes. A new institution – Property Management Bureau (*Wuye guanli ju*) – has been set up to organise the county's rural marketing activities. However, the transition of power has dragged on causing confusion. No such changes have occurred in Dongguan, therefore, this section uses data of 1997 for the purpose of comparison.

not happened. All rural markets have retained their periodic schedule except for Decheng Central Market which operates on a daily basis.

In contrast to Deqing, Dongguan's rural markets have been boosted by the growth of the city's rural economy and high levels of agricultural specialisation and commercialisation. Rapid development of secondary and tertiary industries has not only restructured the latter's rural economy but also reduced the self-sufficiency of peasants. Agricultural specialisation has increased the exchange of products and the level of commercialisation which, in turn, have boosted the number of rural markets.

Dongguan also has a long history of market development. According to historical records, 95 rural markets existed there during the eighteenth century. In 1949, immediately after the Chinese Communist Party (CCP) took over China, there were 58 rural outlets. Rural markets in Dongguan also experienced a marked decline in number during the Great Leap Forward and the Cultural Revolution, by 1963, there were only 22. Unlike Deqing, however, Dongguan's rural markets have undergone remarkable growth during the reform era. Between 1985 and 1995 their number increased from 48 to 189, and by 1997 there were 241. Some designated towns, such as Wangniudun, Changan, Houjie and Xiegang, have more than three rural markets. Moreover, there were more than five market-places in the seat of city government.

The relationship between population increase and market intensification is an important issue that arises in Skinner's (1965a) model of market evolution. He argues that markets intensify simultaneously with population increase. This intensification process is characterised by three contingent steps: (a) an expansion in market size, (b) an increase in market days and, (c) a growth in market numbers (Skinner, 1965a, p.209). An increment in market size is seen as the "most sensitive and virtually automatic" response towards population growth (Skinner, 1965a, p.209). Although the assumptions of this population-market equation are severely criticised by Hodder (1993), this simple formula has been verified in Latin America's and Africa's market development process (Bromley, 1978; Wood, 1978).

Disparity in population growth between Deqing and Dongguan and its impact on rural market numbers seem to support Skinner's argument. Between 1980 and 1997, Deqing's and Dongguan's populations increased 22.5 per cent and 152 per cent respectively (Figure 4.2a) (DGSB, 1995, 1998; DQSB, 1998).[5] The substantial

---

[5] Owing to the deficiencies of the early 1980s data, Dongguan's 152 per cent increase in population is based on the assumption that non-native population (*wai lai ren kou*) accounts for an insignificant proportion in the city's total population in 1985. Concomitantly, the same assumption is applied on Deqing because the same figure is not available.

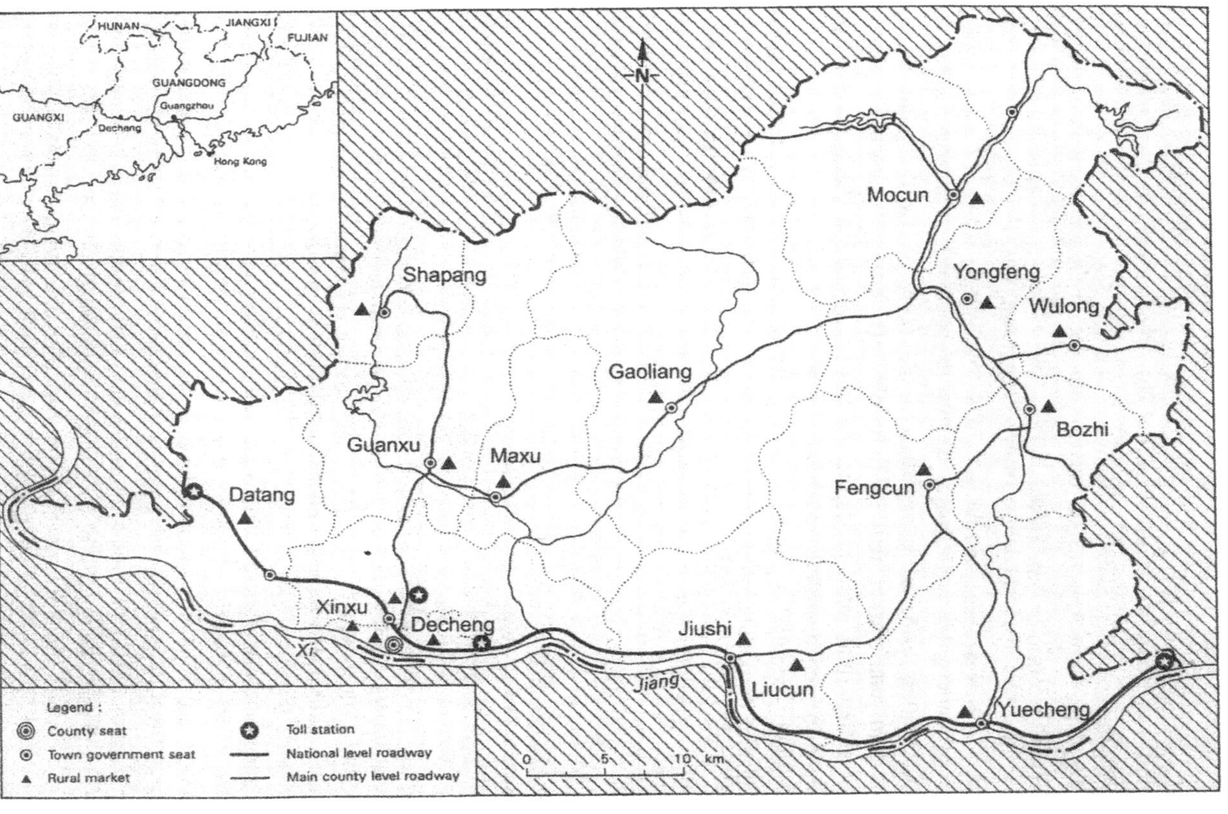

Figure 4.1 Rural markets and settings in Deqing, 1997

variation in the population growth of the two places is associated with markedly different growth rates in the number of markets. Between 1980 and 1997, the number of venues in Deqing increased from 13 to 19, while in Dongguan the expansion was from 35 to 241 (Figure 4.2b). The growth rate for Deqing and Dongguan was 46.2 per cent and 580 per cent respectively.

When the population-market relationship within Deqing itself is examined, however, population growth appears to have little effect on the expansion of rural markets. In 1978, the population of Decheng town was 19,024, and two new markets – Decheng East and West – were established. Until 1997, the town's number of outlets did not change, despite its population reaching 50,000. In contrast, Huilong town had a population of 17,000 in the late 1970s but did not have its own rural market until 1987. In 1997, the local government decided to build a second rural market, despite the town's population being slightly over 20,000 and sparsely distributed in the hilly areas.

Similarly, population growth in Deqing has not led to an increment in market days, despite its population doubling between 1949 and 1997. While Decheng's Central Market has changed to a daily one, the others still remain periodic, and a traditional schedule has been retained. Neighbouring markets have retained different timetables so that peasants can visit other venues without any scheduling conflict. For instance, Decheng, Guanxu and Maxu are three neighbouring market-places within 13 km of each other. Market days in Decheng are on every second, fifth and eighth days of a ten-day cycle, Guanxu on every third, sixth and ninth, and Maxu on every first, fourth and seventh day.

Obviously, market days are busier than ordinary days (Plate 4.1). During market days in Decheng, the pier outside the Central Market is full of 'market vessels' – local boats taking people to market. There is neither a permanent dock nor other boarding facilities at the pier. Passengers carrying pigs and other commodities on shoulder poles get on and off the vessel adroitly over a narrow gangplank, skilful drivers even take their motorbikes on it. Inside the town, the main road is crowded with people, bicycles and tricycles. Most visitors are sun-tanned peasants wearing straw hats, vests and slippers, and carrying small baskets. Some stroll along the street, while others are rushing to the market with piglets, vegetables and fruit on a shoulder pole or bicycle. Children are dressed in tidy clothes and follow their parents happily. Areas nearby are full of street peddlers selling plastic containers, towels, mosquito nets and posters of Hong Kong movie stars and singers. The interior of the market is congested, besides vegetables, there are pigs, puppies, chickens, ducks, vegetables, herbs, sugar cane and bananas for sale. On non-market days all is quiet, there are no 'market vessels' and few peddlers. Many shops are closed except for vegetable and meat sellers. The market-place is empty.

Deqing's low level of agricultural specialisation and commercialisation has not only limited its market days but also trading hours. For example, the busiest period at Guanxu rural market in normal seasons is between 9 am and 10 am.

Figure 4.2 Population, number of markets and turnover value in Deqing

4.2a Population (in millions)

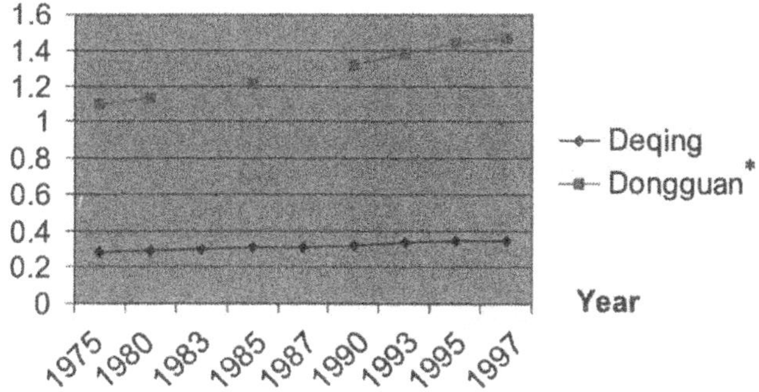

Note: *data of 1983 and 1987 is not available.

4.2b Number of rural markets

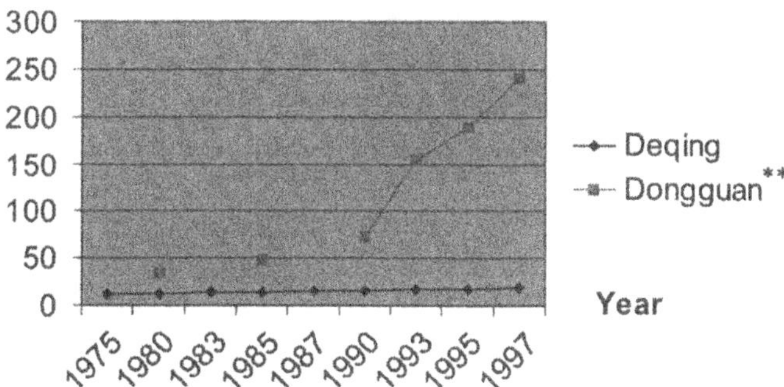

Note: **data of 1975, 1983 and 1987 is not available.

4.2c Market turnover value (in million *yuan*)

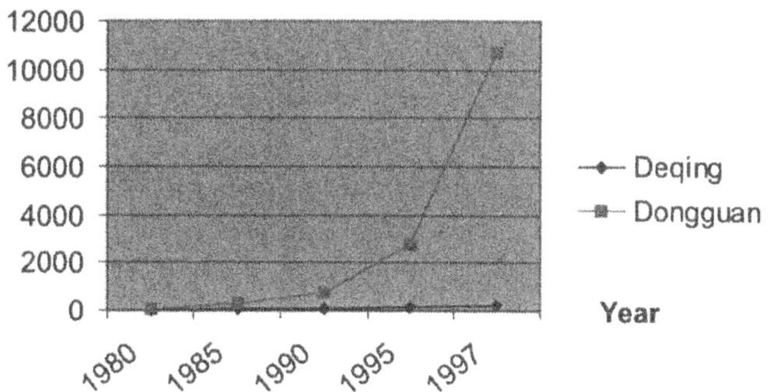

Plate 4.1 Decheng central market during a market day (top) and a non-market day (bottom)

At sowing and harvesting time, the peak hour is delayed by two or three hours because peasants have to finish farm work before they can visit the rural markets. During the farming season, rural markets are less busy. This pattern demonstrates the close relationship between the livelihood of Deqing peasants and rural markets. It also highlights that external influences are minimal in the county.

Table 4.3 Market days at different places in Dongguan city

Name of market	Market days
Wangniudun, Changan, Liubu, Tangxia, Changping, Shipai, Xiegang	Every 1st, 4th and 7th
Dalang, Fenggang, Qishi, Houjie, Zhangmutou	Every 2nd, 5th, 8th, and 10th
Dalingshan, Huangjiang, Qiaotou, Qingqi, Hengli	Every 3rd, 6th and 9th

Source: Fieldwork, and GDPN and GDICMB (1992).

Conversely, economic prosperity has eliminated periodic markets in Dongguan. However, the market schedule still continues as a three market-day in a ten-day cycle (Table 4.3). Since the 1990s, all of Dongguan's market-places have changed to daily operation. New generations have no idea about market days and their past significance in daily life. *Chenxu* ('going to market during market days') is no longer practised and has become a part of history. In Wangniudun town, for instance, every first, fourth and seventh days are still regarded as market days but all now trade on a daily basis. All shops, and meat and vegetable stalls are open every day to provide fresh foodstuffs to local residents. Only peasants are conscious of market days for special agricultural products such as seeds, piglets, fry, and vegetable seeds. The change from periodic to daily markets is an initial result of the intensification process which also shows that Dongguan is one stage ahead of Deqing in this respect.

The rapid expansion in market size also demonstrates that Deqing's market intensification does not completely follow its population growth. Between 1987 and 1997, Deqing's population grew from 310,847 to 340,301 (Deqing County Gazetteer Editorial Committee, 1996, p.699; DQSB, 1998, p.1). Simultaneously, the total area of market-places increased from 38,976 m^2 to 75,700 m^2 between 1987 and 1997 (Deqing County Gazetteer Editorial Committee, 1996, p.458; DQSB, 1999, p.6). The growth rate was 9.5 per cent and 94.2 per cent respectively.

The great increase in market areas was due mainly to the ICMB's intervention rather than the natural result of population growth. During the early 1990s, many venues were redeveloped by the ICMB, and oversized market-places, which did not reflect a town's population size and economic capacity, were built. For example,

Maxu and Yongfeng are two towns with similar features (i.e. same population size and a road links them to other towns). However, since Yongfeng's redevelopment in 1992 and Maxu's in 1995, they have been served by 3483 m^2 and 1129 m^2 market-places respectively. Observation during fieldwork suggests that many stalls in these venues are unoccupied. Moreover, the above figure, which is the only one the government provides, shows the total available market area rather than the real occupied space. This results in an overestimation of Deqing's market size and consequently a misleading assessment of the county's market development.

A similar situation is found in Dongguan. Between 1985 and 1997, the population expanded from 1,208,515 to 1,471,200 and total market area from 81,522 m^2 to 1,614,100 m^2 (DGSB, 1998, pp.181, 319). Thus while the population grew by 22 per cent, the total market area increased by a factor of 19. Again, much of the increment was engendered by the ICMB's improvement projects. Since oversized markets have been built to show off local economic achievements, vacancies are always observed. However, there is 'real' expansion in Dongguan. The emergence of village-level markets, which have been established outside the government system and managed by peasants, also have played an important role in shaping market intensification. Although these village-level outlets are generally small, some of them are over 1,000 m^2 with more than 200 fixed stalls. Examples of these large, village-level markets are those situated in Wusha and Shatou villages of Dongguan.

The variations in population-market growth between Deqing and Dongguan suggest market intensification is not merely a function of economic factors such as population growth and transport. In fact, government intervention plays an important role in restricting market intensification in Deqing. As noted, this is contributed to by the underdevelopment of economic forces under Deqing's semi-closed economy. Strong government intervention constrains rural market development because it results in adherence to the administrative principle of 'one designated town, one market' (see Chapter 3). The consequence of this precept is a slow increase in market numbers, despite population growth being observed. This explains why many towns in Deqing have only one rural market.

The paradox of rapid growth in market turnover and marginal increases in numbers in Deqing is another consequence of government intervention. Between 1978 and 1987, gross market turnover in Deqing and Dongguan increased 42 and 40 times respectively (Figure 4.2c). Given these rates, market growth in Deqing should be faster than in Dongguan as Skinner assumes an expansion in trading volume is a prerequisite for market evolution. However, as noted, only six new markets were established in Deqing between 1980 and 1997, in contrast to an increase of 193 in Dongguan during the same time period.

Relative levels of government regulation and socio-economic development between Deqing and Dongguan demonstrate that different market patterns have occurred in the two places. Rural market density and village-to-market ratios provide a guide to this development. These ratios were compiled using 1997 data

(Table 4.4). There is no point in comparing these measures with Skinner's figures because China under communist rule has created a new social, political and economic context which is very different from his study period. Also, it is not the purpose of this study to estimate the optimal village-to-market ratio. However, it is possible to use other means to demonstrate the substantial differences in market patterns between Deqing and Dongguan. Dongguan's prosperous economy and high population density have contributed to a high market density and a low village-to-market ratio. Conversely, Deqing's semi-closed economy has produced the reverse. There is one rural market for every 10.23 km in Dongguan and one for every 132.75 km in Deqing. Market density in Dongguan is 13 times greater than in Deqing. A circle with a 10 km radius from the city-government seat of Dongguan encloses 12 rural markets. A similar exercise for the county-government seat of Deqing records only three markets. Deqing's closest market-places at Guanxu and Maxu are 7 km apart.

Table 4.4 Market density and village-to-market ratios for Deqing and Dongguan

	Deqing county	Dongguan city
Market density	1:133	1:10
Village-to-market ratio ①	10:1	2:1
Village-to-market ratio ②	94:1	n.a.

Notes: ratio ① is estimated by administrative village and ratio ② by natural village.

Substantial variations in village-to-market ratios between Deqing and Dongguan also illustrate differences in market patterns between the two places. The ratios use the number of administrative villages because information on Dongguan's natural villages is inadequate.[6] The village-to-market ratios of Deqing and Dongguan are 10:1 and 2:1 respectively. This suggests that one market-place serves 10 administrative villages in Deqing and two villages in Dongguan. If Deqing's 1,594 natural villages are taken into account, one market-place serves an average of 94 villages in Deqing. Ratios range from a low of one market per 12

[6] Administrative villages are natural villages with administrative functions. They are unique in China. Local people used to call them "administrative zones" (*xingzhen qu*). An administrative village organises a certain number of natural villages, its size differs between places. In general, an administrative village has a population of about 2,000. For instance, Zhian is one of the 18 administrative villages that are subordinated to Guanxu town in Deqing county. It organises 138 natural villages and has a population of around 1,700.

villages in Decheng town to one market per 232 villages in Gaoliang town (Table 4.5). The only comparable figure for natural villages in Dongguan is Wangniudun town, which has one market for every five villages.[7] This is significantly lower than those in Deqing.

Table 4.5 Village-to-market ratios of different towns in Deqing county

Designated town	Village-to-market ratio*
Decheng	12:1
Xinxu	118:1
Huilong	106:1
Guanxu	146:1
Shapang	75:1
Maxu	121:1
Gaoliang	232:1
Mocun	76:1
Guyou	n.a.
Yongfeng	93:1
Wulong	44:1
Bozhi	89:1
Fengcun	138:1
Yuecheng	162:1
Jiushi	61:1

Note: * is estimated by natural village.

Variations in market size, days, numbers, density and village-to-market ratios underline the fundamental differences in patterns in Deqing and Dongguan. These differences can be highlighted by describing the four stages of rural market development in China identified by Wang (1994):

- *Traditional rural markets*: Peasants have a high level of self-sufficiency. Rural markets are basically places for surplus exchange. Market hinterlands are small and external interactions are rare.
- *Commercialisation of rural markets*: The level of specialisation is increased by an improvement in production. Rural markets become more commercialised and their hinterland expands. Professional traders emerge and rural markets are no longer the place for surplus exchange. Moreover, wholesale and futures markets are formed.

[7] There are 8 rural markets and 42 natural villages in Wangniudun town. Many of the natural villages are large with populations of around 2,000.

- *Modern daily markets and supermarkets*: Commercial and agricultural activities are merged with urban and rural markets. Rural markets are eventually replaced by modern daily markets and supermarkets. Agricultural products are highly commercialised and specialised. Agricultural production is driven by market forces and led by commercial activities.
- *Modern marketing system*: Urban-rural disparities disappear and rural markets are replaced by modern marketing systems, i.e. supermarkets and hypermarkets.

Clearly, Deqing is at the initial stage of Wang's scheme. In addition to the periodic nature of rural markets, evidence of Deqing's primitive system is further derived from its rudimentary hierarchy. As noted, rural markets in Deqing county were initially established under the administrative principle of 'one designated town, one market-place'. Apart from the county seat which has three rural markets and Jiushi town has two, there are 14 designated towns in the county and, accordingly, there are 17 rural markets, including two farm cattle outlets, located at the town level. These rural markets are all located in town-government seats. They are equivalent to the standard markets regarded by Skinner (1964) and form the lowest level of the market hierarchy (Figure 4.3). Decheng Central Market, situated in the county seat and the only daily venue in Deqing, forms the other level – intermediate – of the hierarchy (Plate 4.2). Therefore a relatively flat market hierarchy of two-levels has been created. This structure demonstrates Deqing itself is an intermediate market system – with 16 standard markets dependent on it. Moreover, since the system has been established under administrative principles, its economic hinterland is bound by Deqing's administrative boundary, i.e. 2,257 km. Decheng town, the county seat, serves as both economic and administrative centre of this local system.

All markets in Deqing's hierarchy are at the town-level, though some are large and others small (Plate 4.3). Among the former, Gaoliang town has 550,000 visitors per year and a turnover value of 132.7 million *yuan*. Ranged after Gaoliang in 1997 were Bozhi with a turnover of 131 million *yuan* and Maxu with 127.5 million *yuan*. Conversely, Shapang, like Datang and Liucun, is a small market with 40 stalls, 180,000 visitors per year and a turnover of 2.4 million *yuan*.

Despite local distinctions, there are no differences in functions between large and small markets. All are engaged in exchanging surpluses and retailing basic necessities such as vegetables, pork, household utensils, and clothes. As commercial activities are not as widespread as in the Delta, no wholesale markets have been established. Although 40 stalls are assigned for clothes in Guanxu, less than one-quarter are leased. Many are selling foodstuffs, such as vegetables, pork and poultry, and agricultural related products, normally on short-term contracts of

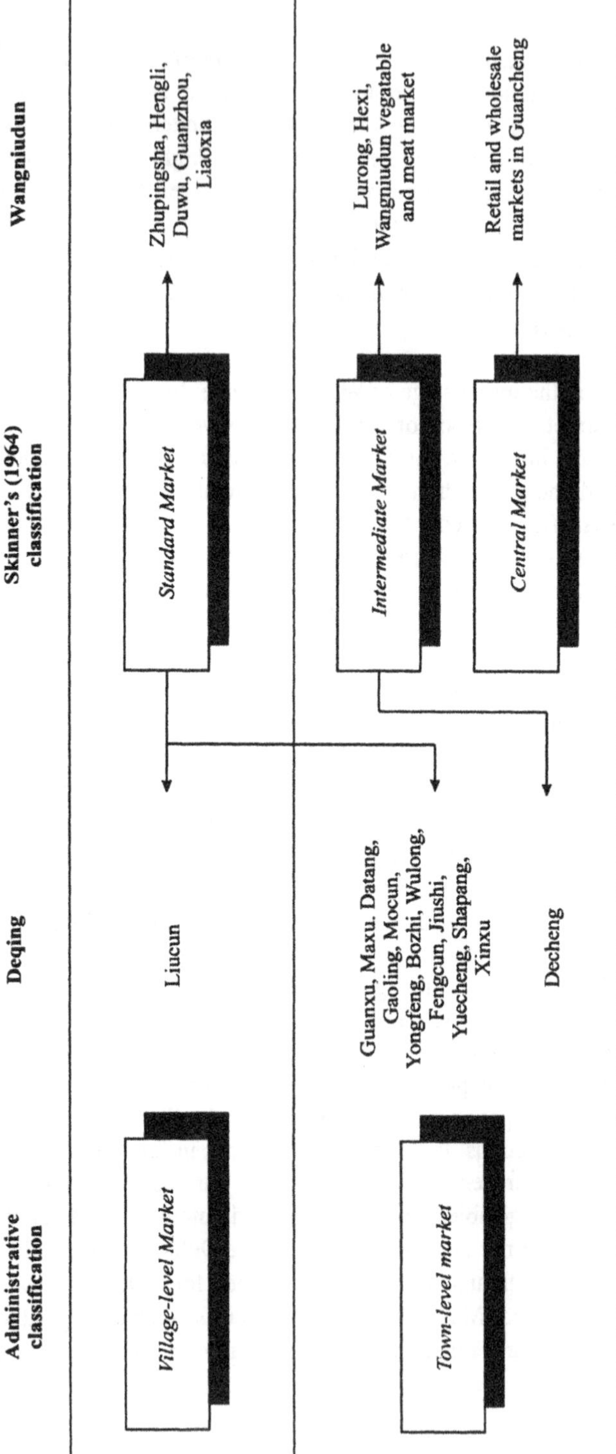

Figure 4.3 Market classification in Deqing and Wangniudun, 1997

Plate 4.2 Decheng central market (an intermediate market in Deqing)

Plate 4.3 A town-level market (standard market) in Deqing

one to three months. Such a leasing situation reflects the spasmodic nature of agricultural surpluses.

The only village-level outlet in Deqing is located in Liucun of Jiushi town – a traditional site since Imperial times. It was designated as a rural market by ICMB in 1974. This small market has only 48 temporary stalls and 21,900 visitors per year, facilities are primitive and old, and it is the sole rural market in Deqing still practising street peddling. Elsewhere in Deqing, small shops which sell groceries, salt, cigarettes, candies, wine, and other daily necessities are scattered in the county's 1,584 natural villages. In some remote areas, there are mobile traders on motorcycles selling pork. These rural shops and mobile traders, however, are not regarded as formal agents in the county's retail system because they are operated by peasants on a part-time basis outside farming schedules. There are no official figures of their numbers, particularly as some of them are unlicensed.

Three markets in Decheng town – the county seat – form the top level of Deqing's market hierarchy. They are all located along the town's main road, Zhaoyang road. Improved in 1992, the two-storeyed Decheng Central Market is the largest in the county. It is classified as 'comprehensive' with a total area of 7,266 m^2. The ground floor is designated for processed and fresh foodstuffs and the second floor is for 'dry commodities' such as clothes, shoes and other household articles. Inside there are 506 fixed stalls which attract 2 million customers each year. A rural market is located behind the complex and is about 50 m from the town's pier, where all 'market vessels' stop during trading days. There are both fixed and temporary stalls in the rural market. Three hundred metres to the east of the Central Market is Decheng Eastern Market, a comprehensive one with 108 stalls. The Western Market is situated 500 m to the west of the Central Market, and specialises in seeds but foodstuffs are also available. The scale of these two markets is smaller than that of the central one. Local people used to call them the 'Decheng markets'.

There are substantial differences between the functions of Decheng and other town-level markets. Although exchanging surpluses and retailing are still major activities, clothing and daily necessities sold in the Central Market are high-order goods. Small electrical appliance shops are also found. While stalls selling foodstuffs account for about one-third of the market, more than half of the stalls are selling clothes, shoes and other accessories. More variety is available therefore. However, despite these features, the function of absorbing surplus rural labour is not as obvious as in the Delta area because many peasants still have self-sufficient lifestyles. Moreover, market development remains embryonic because some stalls and shops are owned by peasants who still depend on farmland to survive and regard trading as a part-time activity. All of these characteristics demonstrate that the rural market system in Deqing has yet to be modernised.

Market structures and hierarchies within Dongguan are relatively complex. Although Dongguan's rural markets were initially established according to the administrative principle of 'one designated town, one market-place', the rapid development of TVEs and the village-level economy have stimulated their growth in villages. Besides the 32 rural markets in designated towns, many more have been established in Dongguan's 551 administrative villages. In addition to the retail and wholesale outlets in the urban district of Dongguan city, a three-level market hierarchy has been created.

Unlike Deqing, rural markets at the village level form the lowest level – standard market – of Dongguan's hierarchy (Figure 4.3). Village markets have been established by administrative villages but not by the local governments or ICMB; they are usually small in scale and open daily. For example, Duwu administrative village within Wangniudun town has one of the five village-level markets there (Figure 4.4). It has only 20 stalls but opens every day and all facilities are in permanent concrete structures (Plate 4.4). Due to the village's significant agricultural specialisation, the market is not designed to sell surpluses but to supply daily foodstuffs such as fish, pork, vegetables and other non-staple foods. According to local cadres, the prime reason for establishing a village market is to provide convenience for local residents. Most customers, therefore, are residents of the administrative village. As the peddlers are former peasants, the rural market has absorbed surplus labour.

Rural outlets at the town-level form the intermediate level of Dongguan's market hierarchy. Unlike village markets, they are formally established and operated by the ICMB. Situated at traditional sites, they changed to daily operations during the 1990s. Although market days have been retained, they are no longer relevant. Wangniudun town has three town-level market-places and all of them are situated in the town-government seat (Figure 4.4). The foodstuff outlet was established in 1985 and has 158 fixed stalls selling pork, fish, vegetables and other non-staple foods to town residents (Plate 4.5). In 1989, a modern three-storey trading complex was built, with an area of 7,500 m^2. The first floor is given over to foodstuffs and has 94 stalls. The second and third floors are used for clothing, footwear, household appliances and other necessities. Restaurants and cafés are also found on the third floor. In 1993, a bigger complex – known as Hexi market – was established in the town-government seat; this has not only provided more shops and stalls, but also a western style supermarket which has attracted many visitors. As in Wangniudun, the market-place has been transformed into a modern shopping centre.

Wangniudun's case demonstrates that an intermediate market system has formed at the Dongguan town level, while a similar structure is found at the Deqing county level. Since this system was initially established by administrative principles, as noted, its economic hinterland is basically bound by Wangniudun's administrative boundary, i.e. 30 km^2. By comparison with Deqing's 2,257 km^2, Wangniudun's market hinterland is 75 times smaller than the same intermediate market system in a hilly area. Given all the socio-economic and administrative

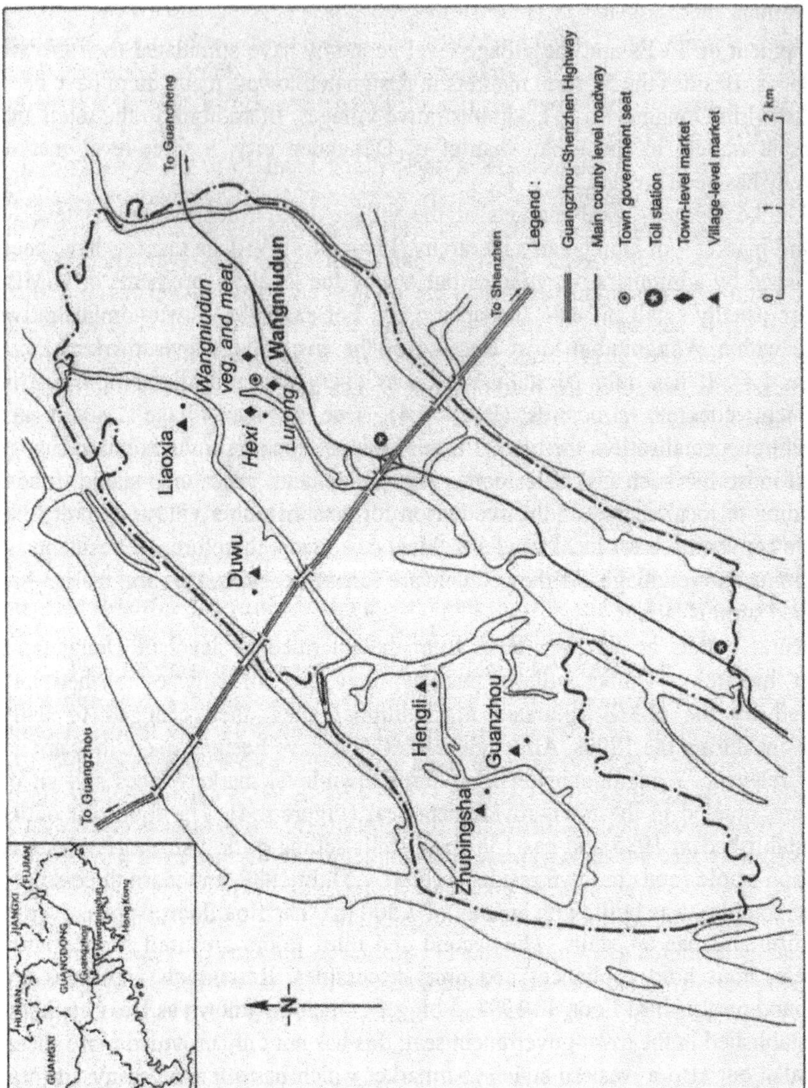

Figure 4.4 Rural market system in Wangniudun town, Dongguan, 1997

Plate 4.4 A village-level market (standard market) in Wangniudun

Plate 4.5 Wangniudun market situated in the town-government seat

changes since Imperial China, these figures cannot be compared with Skinner's (1965a) estimation (see Chapter 2). However, the substantial difference between Wangniudun and Deqing generally supports Skinner's (1965a) argument that the marketing area of an intermediate system is small in plains but large in hilly/mountainous areas.

The large-scale markets in Guancheng, the city government seat, are at the highest level of Dongguan's hierarchy. They are equivalent to the central markets noted by Skinner (1964). Of the seven situated in the city district, the three-storey Kehu market is in the city centre with a total area of 6,000 m^2. A non-staple food wholesale market has also been established on the ground floor of the complex. Kehu market is one of the busiest and each day has between 6,000 and 8,000 customers. Not far away from Kehu is Qincaitang market, established in 1979, and the largest in the city government seat. It has a total area of 11,300 m^2 and over 270 permanent stalls/shops. The market provides products ranging from foodstuffs to household articles, cultural and educational items, clothes and footwear, and electrical appliances. A fast food eatery, restaurant and small department store are also present. Over 10,000 people visit the complex each day.

There are also wholesale markets at Dongguan's highest market hierarchy. However, Skinner (1964) does not address these in his thesis because they were rarely found in the countryside during Imperial China. During the mid-1980s, wholesale markets for fruit and aquatic products were also established in the city government seat. Locally grown fruit, such as bananas, sugar cane and lychees, have been exported from here, and fruit from northern China has been imported. The aquatic products market attracts specialised peasants in designated towns to exchange products. Over 1,000 *jin* are traded daily. The presence of wholesale markets demonstrates that the process of market modernisation has taken place in Dongguan.

Thus, Deqing's rural market development is still at an initial stage. However, the emergence of daily markets in the county seat and an increase in specialisation levels indicate progress in moving from a traditional to a modern stage. Yet this is constrained by Deqing's low level of economic development which is inferior to that of Dongguan. In contrast, a more complex and advanced market system has been formed in Dongguan. Moreover, when old standard markets (town-level ones) become intermediate, new standard markets (village-level ones) emerge. The modernisation process as suggested by Skinner (1965a) is observed. Despite these features, the system in Dongguan has not reached the most sophisticated stage outlined by Wang (1994) because of persistent disparities between urban and rural areas, and within rural areas. Small peasant farmers still exist, although agricultural activities are highly commercialised.

The relatively advanced rural market system in Dongguan implies that the city has gone through a modernisation process during the reform era. However, its market intensification is a top-down affair rather than a bottom up one as described by Skinner (1965a).[8] Dongguan's top-down movement has involved the use of administrative parameters such as the licence system, administrative control over the establishment of markets, and local government intervention. Although the emergence of village markets in Dongguan demonstrates that marketing activities are driven more by economic forces than administrative parameters, the persistence of government intervention has not given free rein to market forces. Accordingly, Dongguan's village-level markets can only be seen as hybrid structures derived from the interplay of economic forces and administrative parameters. This hybrid system is the unique product of the transformation of China's economy. A market system created by economic forces has yet to appear.

In short, the developmental gap between market systems in Deqing and Dongguan is substantial. Rural market development is not only slow in Deqing, but market density is low and this contributes to the county's very high village-to-market ratio. Moreover, the two-level hierarchy shows that the system in Deqing is primitive and rudimentary. In contrast, rural markets in Dongguan have flourished during the reform era. Market density is high and each market services a relatively small area compared to its counterpart in Deqing. Rural markets have also been extended to the village-level and form the base level of the city's market hierarchy. In addition to the establishment of wholesale markets, a more advanced system has been created. Although development is well ahead of its Deqing counterpart, the system in Dongguan is still far removed from a completed rural market system comprising retail, wholesale and futures markets (Wang, 1994). However, given the high concentration of industrial development in Dongguan, it is doubtful whether any agricultural futures markets will develop. For the same reason, a complete rural market system might only be found in China's inland areas but not in the country's coastal areas.

Despite the disparities in rural market development, the case of Deqing and Dongguan demonstrates that their economic and administrative hierarchies overlap. All rural markets are situated in either town-government or county seats, and those at village-level in Wangniudun – established by economic forces – are no exception to this rule. They are located in village committees, i.e. administrative central places at village levels. Deqing and Dongguan are not exceptions. They reflect the

---

[8] The top-down process of market development is not unique to China. It has also been found in Puno, Peru and the Changhua plain of Taiwan (Appleby, 1976; Crissman, 1976a).

relationship between economic and administrative hierarchies during the reform era. Although the political-cum-economic structure has persisted, one should not automatically assume that the degree of control, or government regulation, is equally high in both Deqing and Dongguan. This is demonstrated in subsequent chapters.

Rural market development in Deqing and Dongguan does not follow any formula. It is, however, closely aligned to the changing local context of the respective area. The two places have varied with regard to foreign investment input, industrial and agricultural development, commercial growth, accessibility, and level of liberalisation under different reform policies. These variations have created very different market systems in Deqing and Dongguan – the former is still primitive and rudimentary whereas the latter is advanced and complex – despite both being highly integrated with their respective administrative hierarchies. Moreover, as market-places are regarded as "constantly changing streams of people and good" (Hodder, 1993, p.32), their development is affected by a variety of factors over time and space. As a result, Skinner's plains and hilly/mountainous models in term of spatial variations are too primitive to provide any satisfactory explanation on rural marketing. Apart from the physical landscape, there are many economic and social elements that generate spatial differences. The emergence of these components, such as rural industrialisation and foreign investment, demonstrate that the dynamics of economic and social conditions vary over time and space – an observation weakening the contemporary explanatory power of Skinner's models.

The comparison of Deqing and Dongguan shows market intensification is not necessarily generated by endogenous processes as suggested by Skinner. While administrative principles – an example of internal forces – build the skeleton of a rural marketing system, long-distance trade and specialisation improve its development. In Deqing, the lack of external connections has generated a rudimentary market network with stagnant commercial activities. Conversely, transregional commercial activities have changed Dongguan's marketing system into an advanced one. Indeed, the critical impact of long-distance trade on promoting local markets has been demonstrated in Europe, Africa and China in ancient times (Good, 1973; Rowlands, 1973; Spencer, 1940). These instances, together with the contemporary example of Deqing and Dongguan, echo Hodder's (1993, p.50) argument that "trade and market are by definition exogenous phenomena". Although it is possible that endogenous forces, such as population growth, may lead to market growth, overlooking external influences has created a major deficiency in Skinner's thesis.

Chapter 5

Beyond the Preoccupation of Transport: Economic Factors and Their Impact on China's Rural Market Development

This chapter examines locality effects and their impact upon the marketing patterns of Deqing and Dongguan. Locality effects include the level of self-sufficiency and incomes of peasants, the degree of commercialisation, road routes, and administrative principles. These factors and how they shaped rural market intensification and modernisation were addressed in Skinner's (1964, 1965a) sophisticated framework. While they remain critical, their significance has shifted in response to changes in China's economic, social and political conditions since the Imperial period. Specifically, this chapter investigates the interactions between peasants and rural markets in Deqing and Dongguan. It establishes the significance of peasants in shaping rural market development in their roles as both consumers and producers, and how they provide the dynamics for growth. The focus is on economic factors, while non-economic ones – the establishment of markets and government regulations – are discussed in the next chapter.

Self-production or market purchase: The level of self-sufficiency and its impact on rural market development

The level of self-sufficiency crucially affects China's rural market development. It is a critical indicator of the interrelationship between the production patterns of a specific place and its economic activities. According to Skinner (1965a, pp.208-209; 212-213), a decrease in household self-sufficiency increases food demands and thus stimulates market intensification and modernisation. He does not disclose the underlying causes of the decline in self-sufficiency, but assumes it is a 'natural' process driven by increases in agricultural productivity. Generally, his argument reflects the correlation between agricultural production and rural market development in Imperial China which was characterised by a closed economic system with minimum regulations. Although fieldwork in Deqing and Dongguan supported the thrust of Skinner's observations, the level of self-sufficiency is not necessarily a 'natural' process and many 'artificial' factors are noted.

One of the 'artificial' factors which affected the level of household self-sufficiency during the reform era is the heavy tax and improper levies (*luan shoufei*) impost – a result of institutional corruption (Wedeman, 1997). These levies have led to the persistence of 'high self-sufficiency and low market interactions'. This situation is illustrated by reference to Mr Lu.

> Mr Lu lives in Shengtian village of Huilong town and cultivates 4 *mu* of paddy field which, in 1997, yielded 4,000 *jin*. Despite the relatively high yield, Mr Lu had never thought of selling his grain on the rural market because he had to reserve it for taxes-in-kind. Tax items include an agricultural tax, administration fees for the town government, collective allotted payment, water conservation tax and others. The amount of tax is set in accordance with the number of people in his household, and the area of the paddy field owned by Mr Lu. The amount is estimated in monetary terms and then converted to grain. Taxes can be paid in either cash or grain, or both. Many peasants pay their taxes in grain because they rarely have a cash income.
>
> In 1997, Mr Lu paid a total of 2,498 *jin* of grain for tax (Table 5.1). Given that Mr Lu's yield was 4,000 *jin* in 1997, taxes accounted for more than half of the paddy yield. The remainder was retained by Mr Lu as his family's food grain for the year, with little surplus remaining for sale. Mr Lu said tax and all minor levies were the major reason for this situation. Moreover, as Mr Lu's paddy is situated in the hilly area of Deqing, there is little he could do to increase the harvest. Occasionally, bad weather and natural disasters have reduced yields. Despite these occurrences, tax items and payments have increased annually. During poor years, tax payments were in arrears. Conversely, in normal years, the amount of grain after tax was just enough to feed his five family members.

According to Mr Lu, peasants only sell their grain when they need money for emergency uses such as paying children's tuition fees and other local levies. By retaining the grain for tax and self-consumption, peasants minimised the need to sell it in rural markets. Thus rural market activities remain stagnant, as the major function of surplus exchange is undermined.

Mr Ou's case is worse than that of Mr Lu:

> Living in Shangpao village of Huilong town, Mr Ou had 3 *mu* paddy field to support his family. Since Mr Ou's field comprised several small ones scattered over different areas, the yield was low. In 1997, his total harvest was 2,400 *jin*. Tax payments had accounted for over half of the total yield. Mr Ou also said the tax burden had increased year by year (Table 5.2). In 1995, when Mr Ou had 5 *mu* paddy field, the tax payment was 1,534 *jin*. In 1997, after he lost 2 *mu* field, the tax was 1,481 *jin*. As a large proportion of the yield was given to the government for tax, the remainder was not enough

to support his four family members. This situation put Mr Ou in debt. Consequently, Mr Ou's daughter had left home and worked as a migrant worker to support the family. Mr Ou's son, who had just graduated from junior high school, was also planning to look for a job in the Delta area.

Table 5.1 Mr Lu's tax assignment (in grain) in 1997

	Items	Grain to submitted (in *jin*)
1	Agricultural tax (in grain)	212
2	Agricultural tax balance due	--
3	Grain procurement	654
4	Grain procurement balance due	577
5	Administrative fee for town government	627
6	Collective allotted payment	60
7	Water conservation grain	120
8	Civilise district construction grain	20
9	Land rent	56
10	Last year's balance due	172
	Total:	2,498

Table 5.2 Mr Ou's tax assignment (in grain) in 1995 and 1997

	Tax items	1995*	1997**
1	Agricultural tax in grain	277	277
2	Agricultural tax balance due	294	--
3	Purchase grain	730	680
4	Administrative fee for town government balance due	--	85
5	Education payment (in grain)	50	110
6	Water conservation payment (in grain)	69	138
7	Land rent	37	110
8	Collective allotted payment	37	--
9	Civilise district construction	--	20
10	Others	40	61
	Total	1,534	1,481

Note: * based on 5 *mu* paddy field; ** based on 3 *mu* paddy field.

With taxes absorbing the surplus grain of peasants, they only had sideline products to sell in markets. Recall Mr Lu's case:

> In 1997, the two pigs Mr Lu owned were the only products he sold at the market (Figure 5.1a). As a producer, this was the only interaction Mr Lu had with rural markets. Beside that, he rarely visited markets despite the closest venue – Datang market – being situated only 3km from his village (Figure 5.1a).

In Mr Ou's case:

> He did not keep pigs but was involved in pine resin production during the slack season. Pine resin was sold at a fixed price to authorised agents at designated purchase points and cash was paid. Although the procurement of pine resin was monopolised by the local government, selling pine resin was the only 'commercial' interaction Mr Ou had negotiated with the market (Figure 5.1a).

The above discussion demonstrates that the county's embryonic market development has, in turn, enforced high levels of self-sufficiency. Stagnant trading activities result in difficulties in selling surpluses.[1] Moreover, low surpluses affect the consumption levels of peasants because sale of excess grain is the only major source of income in a semi-closed economy like Deqing.

Conversely, in Dongguan, taxes and levies do not have the same negative effects. Take Mr Dui's case as an illustration:

> Although Mr Dui, like his Deqing counterpart, has to pay an agricultural tax and a number of levies to the government, these are supported by the collective unit – the administrative village – as a collective economy flourishes in Dongguan. Unlike Deqing, such a tax is paid in cash rather than in grain. As Mr Dui did not personally take care of taxes, he did not know how much he paid each year. His largest expenditures were in production costs such as the field-subcontracting fee, purchasing farm chemicals, maintaining the irrigation system and other farming facilities. Despite these, Mr Dui earned a stable income from specialised farming. Although he did not disclose his total income, he said it was adequate to

[1] See Mei (1998) for an analysis of this situation.

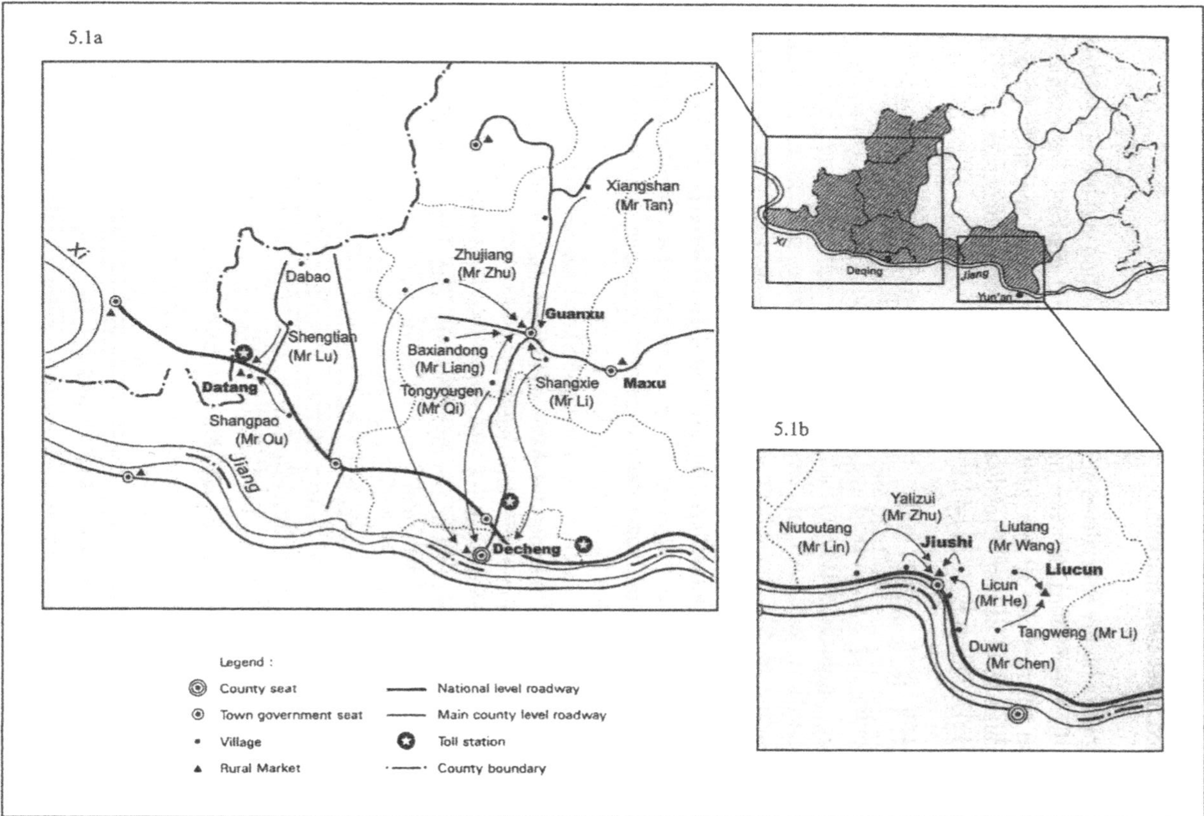

Figure 5.1 Peasant-market interactions under the impact of high self-sufficiency and low incomes in Deqing, 1997

> support his five family members, and also enabled him to buy everything he needed in the rural markets.

Dongguan's highly specialised agricultural production and low tax burden has intensified the interactions between peasants and rural markets. For example, in Wangniudun town, over 50 per cent of the peasants are specialist farmers – implying a high level of commercialisation in agricultural products as all are sold in rural markets. This is typified by Mr Dui:

> Mr Dui is a typical specialised farmer in Duwuxin village. While subcontracting a 4 *mu* field and a 12 *mu* fish pond from the collective unit, Mr Dui produced 800 *jin* of fish, 13,000 *jin* of sugar cane and 15,000 *jin* of bananas last year. All of these products were sold in rural markets. According to Mr Dui, about 40 per cent of the fish yield was sold to wholesalers from other places, the remainder being sold to local retailers in Wangniudun rural markets. As Mr Dui farmed several species of fish, maturing at different times of the year, he had fish for sale almost every two months. Bananas and sugar cane were also sold once or twice annually to both wholesalers and local retailers.

Frequent peasant-market interactions have stimulated market activities and thus generated a rapid expansion of the rural market system.

Government land requisition has effectively weakened rather than reinforced the self-sufficiency of peasants in Deqing. However, this does not imply that the requisition of land has fundamentally altered production patterns of peasants or accelerated the county's extent of specialisation and commercialisation. Nor has it increased the number of non-agricultural positions for peasants. In fact, farmlands have been taken over for road construction so peasants can no longer depend on their land to survive. As they have lost their means of production (i.e. land), they have turned to rural markets for food. Such changes have intensified interactions between peasants and markets and provided a major force for further rural market development.

Much land requisition has occurred in villages located along the new National Roadway No.321 in Jiushi town (hereafter abbreviated as the 'New Road 321'). Take Mr Chen of Dúwu village as a case in point:[2]

> Land requisition resulted in Mr Chen losing his private plots and 2 *mu* of paddy field, leaving him with less than 1 *mu* of paddy field to cultivate. Consequently, Mr Chen's annual yield was insufficient to support him and

[2] Dúwu (with accent) is in Deqing; Duwu (without accent) is in Wangniudun, Dongguan.

> his four family members. According to Mr Chen, he rarely visited rural markets before the land requisition occurred, although Jiushi market is located only 2 km from Dúwu village (Figure 5.1b). Loss of farmland, however, had pushed Mr Chen to visit rural markets every two days for grain and other foodstuffs (Figure 5.1b). Thus, his connections with rural markets had become more frequent. Cash compensation from the requisition of his land allowed Mr Chen to effect a smooth transition from high self-sufficiency to buying from the market. This change has eroded Deqing's vicious circle of 'high self-sufficiency stagnant market development' by pushing peasants to the market for the food they need.

As peasant-market interactions have intensified, the rural markets' function of commodity exchange has been reinforced.

However, in Deqing the benefits of requisitioning land have been wiped out on occasions due to the embezzlement of compensation by the local authority. This situation provides a counter argument to Skinner's suggestion that a decrease in self-sufficiency inevitably increases demand and market growth. For example, Yalizui is a medium-sized village of 600 people, or 110 households, located along the New Road 321. Construction of the national roadway took over 60 *mu* of paddy land and dry farmland from the village. Compensation for the requisitioned paddy fields was 3,000 *yuan* per *mu*, and for dry farmland 1,500 *yuan* per *mu*, paid collectively to the administrative village (*xingzheng cun*) for disbursement to natural villages (*ziran cun*).[3] After deducting administrative fees and many other levies, less than half of the proceeds reached the dispossessed. When peasants do not have money to spend, rural market development is affected as the dynamic for growth is depressed.

> Mr Zhu's family is one of the households in Yalizui village to suffer from inadequate compensation from land requisition (Figure 5.1b). Mr Zhu had 5 *mu* of paddy field before land requisition. The usual annual paddy yield not only provided sufficient grain to support his family, but also surplus for occasional sales. Road construction took away more than 4 *mu* of farmland from Mr Zhu. Now cultivating less than 1 *mu* farmland, Mr Zhu was worried about his future because he would not be able to produce enough

---

[3] In China, villages can be classified into administrative and natural villages. The former are village-level administrative units which local people used to call 'administrative zones' and 'production brigades' during the post and pre-reform periods respectively. Natural villages were known as 'production teams' before 1979. Instead of village, many peasants still use this term to refer to their village, and other neighbouring villages.

> grain to support his six family members. Nor did the government's minimal compensation enable him to make the transition from self-sufficiency to a dependence on markets. He stressed that the meagre compensation was not enough to make up for the paddy field he lost and the stable life he once had. Mr Zhu's dilemma was that while he could not maintain self-sufficiency, he did not have money to purchase rice and other foodstuffs in the market. Accordingly, Mr Zhu's son has to work outside the village to supplement the family income. Mr Zhu remembered clearly that 1998 had been the third year he had bought rice in the local rural market. While land requisition had intensified Mr Zhu's connections with rural markets, insufficient compensation choked off these tentative linkages. Consequently, Mr Zhu struggled to survive because purchasing commodities in markets requires money.

In this case, increased peasant visits did not greatly affect the 'high self-sufficiency low market interactions' correlation.

The embezzlement of compensation creates obstructions to rural market development. It prevents interactions between peasants and rural markets. Although land requisition provides peasants with a golden opportunity to abandon a self-sufficient livelihood, many think that local government and developers have taken away their farmland and destroyed their self-sufficient livelihood. One of the peasants to share this view is Mr Lin.

> After losing 4 *mu* of paddy field, Mr Lin had only 2 *mu* of farmland left to support the family. Unlike Yalizui village, compensation has not been paid to Niutoutang village peasants. This has not only aggravated Mr Lin's strong resentment towards local government, but also left him in immediate financial difficulties. Although Mr Lin's paddy land had a high yield of 500 *jin* per *mu*, this was not enough to support his seven-member family. Consequently, like many peasants located along the New Road 321, Mr Lin has had to purchase grain from the local market with his tiny income from pine resin production (Figure 5.1b). Still, he did not have a stable income. According to Mr Lin, his living standard had deteriorated with the increase in market interactions rather than improved because he had to spend money in markets.

Unlike Deqing, land requisition in Wangniudun town of Dongguan has stimulated market growth, as those dispossessed have transferred to non-agricultural activities. Interactions with markets have thus intensified because peasants are now dependent on them rather than farmland for grain and foodstuffs. For example, rapid economic growth has converted Hedong village's 300 *mu* farmland into an urban landscape since the early 1990s. With compensation – including cash and grain – being paid to all those dispossessed, many peasants used the money to develop small businesses.

> Mr He started a passenger delivery business and became completely divorced from the agricultural sector. This change has resulted in Mr He visiting rural outlets for grain, non-staple foods, clothes, footwear and all daily necessities. According to him, he visited rural markets almost every day and sometimes different venues for different products. Mr He is not an exception, as all the villagers in Hedong are reliant on markets for goods.

Transferring peasants from agricultural to non-agricultural activities has consolidated Dongguan's 'low self-sufficiency and high market interactions' correlation which, in turn, has stimulated rural market development as more people are dependent on them for survival.

Natural disasters have also weakened the 'high self-sufficiency and low market interaction' of peasants in Deqing. This factor is considered as artificial, or semi-artificial, because it is contributed to by Deqing people's philosophy of: 'living on a mountain, live off the mountain'. When natural disasters occur, peasants are pushed to the market for foodstuffs. Again, in many cases, this has occurred in villages found along the New Road 321 in Jiushi town. The new roadway, which runs along the side of the Xijiang River, is much higher than the water level. According to the Roadway Bureau, the roadway was designed to avoid it becoming impassable during flooding, it also acted as a dyke to protect the villages and farmlands. However, the roadway's poor drainage design has prevented inland waterways from flowing out to the Xijiang River, and flooding has occurred inside the dyke during the rainy reason. Consequently, many paddy fields were submerged, causing serious loss. Tangweng, a medium size village with a population of 500, has suffered from inland flooding. Since the opening of the New Road 321, floodwater has submerged the fields for several months each year. As a result, there is only one crop in Tangweng village.

> Despite Mr Li having 1.6 *mu* field to support his family of two adults and two children, he only produced enough grain for 10 months, so he resorted to purchasing rice from rural markets. Although Tangweng village is situated 3 km to the south of Liucun market and 5 km to the north of Jiushi market, Mr Li rarely visited these venues before the onset of regular flooding (Figure 5.1b). The need to purchase supplementary grain has resulted in him visiting rural markets more frequently than before.

Such a change is significant for rural market development because peasants increasingly have to depend on markets for foodstuffs.

Inland flooding in Jiushi also benefits market development by changing the level of self-sufficiency of peasants. For example:

> Mr Wang, who lives in Liutang village and cultivates 4 *mu* paddy field, had seen his annual yields significantly decreased by inland flooding. After deductions for tax, the annual yield of 300 *jin* per *mu* became insufficient to support Mr Wang and his four family members. In recent years, he has not only visited markets more frequently, but has extended sideline production to increase his income because market trading requires higher capital resources. Like many Deqing peasants, Mr Wang rarely visited markets when he was self-sufficient, although Liutang village is located only 2.5 km from both Jiushi and Liucun rural markets (Figure 5.1b). While buying supplementary grain from rural markets, Mr Wang kept 5 pigs and hoped to sell them at a higher price to increase family income. He also worked part time during slack seasons. Thus flooding has altered Mr Wang's high level of self-sufficiency by changing his production pattern and intensifying his interactions with markets.

Such changes are undoubtedly important factors pushing market development in Deqing.

The two problems of land requisition and flooding in Jiushi town have clearly helped to produce a closer relationship between peasants and rural markets. Where these have not occurred, paddy yields are generally adequate for tax and self-consumption. In these places, peasants have a very high level of self-sufficiency and never buy grain from markets, although occasionally they do purchase non-staple foods and clothes. Discussions with Deqing peasants suggest that self-sufficiency is their preferred lifestyle, many think that having to pay for grain is a sign of poverty. This perception has limited Deqing's rural market growth, despite interactions between peasants and rural markets being intensified.

The variations in peasant-market correlation between Deqing and Dongguan can also be viewed from the perspective of peasants as consumers. As peasants generally consume their own grain and vegetables in Deqing, they rarely purchase these staple foods in rural markets.

> Mr He, who lives in Licun village and cultivates 2 *mu* farmland, illustrates this well. As with many Deqing peasants, all grain and vegetables were supplied by his own farm. Consequently, there was little need for him to buy things in rural markets, apart from occasional visits to Jiushi rural market for salt, soy sauce, sugar and clothes, although Liucun rural market is located only 2 km from his village (Figure 5.1b). Mr He did not remember exactly how many times he had visited rural markets in the last year but he thought it was less than ten.

His interaction with rural markets is constrained therefore by his self-sufficiency.

Given the semi-closed economy of Deqing county, there are very few internal economic forces to counter the high level of self-sufficiency. Accordingly, weak interactions between peasants and rural markets have persisted. Moreover, a high level of self-sufficiency also constrains the cash income of peasants. In Mr He's case, however, he had a small supplementary income from non-agricultural activities – the collection of pine resin. Nevertheless, like many peasants in Deqing, he was still dependent on land for survival, and work on the farm was his family's primary activity. Limited cash income also constrained consumption by peasants in markets. This also explains why rural market development in Deqing has been slow and the market system remains rudimentary.

Conversely, the high level of specialisation (i.e. low sufficiency level) has consolidated Dongguan's 'low self-sufficiency and high market interaction' pattern. Not only do peasants sell their farm products in markets, but they also buy almost everything they need there. In fact, self-sufficiency has been consigned to history in Dongguan, with villagers, as consumers, depending on rural markets for everything they need. Accordingly, frequent connections between peasants and rural markets are observed. For example:

> Mr Du is a specialised farmer who lives in Duwu village of Wangniudun town. Over 80 per cent of his farmland was used for growing bananas, one of the most popular cash crops in Dongguan. Unlike his Deqing counterparts, Mr Du did not grow either paddy or vegetables for his own consumption. All foodstuffs, including staple and non-staple foods, were purchased from the rural market. Mr Du did not mention how much grain his family used each year or the cost. However, he reported the grain price had been very stable in recent years and his income allowed him to buy grain in markets. The opening of a village-level market in Duwu was very convenient for him. He visited it almost daily for pork and vegetables. Besides foodstuffs, Mr Du also bought clothes, footwear, children's apparel, farming tools, farm chemicals and other household articles in the Hexi market, the town-level market located in the town-government seat (Figure 5.2). According to him, he visited the rural market three to four times per week, although he did not necessarily buy goods every time.

Mr Du's case is no exception. Ninety per cent of Duwu village's farmland is taken up by the growing of economic crops such as bananas, sugar cane and vegetables. As many peasants are specialised farmers, they make money from their cash crops and buy grain and non-staple foods in rural markets. So commodity exchange has become an important channel to obtain the goods they need, and their interaction with rural markets is extensive; a fact which also explains why a village-level market was established in Duwu village but not in Deqing.

Thus, the level of self-sufficiency is an important factor influencing rural market development in Deqing and Dongguan. As producers, the tax burden of peasants, land requisition, and flooding are three components affecting the levels of self-sufficiency of peasants and their degree of interactions with rural markets. These factors are considered as 'artificial' because they are caused by local government policy and the life style of Deqing people. While the tax burden of peasants has reinforced Deqing's 'high self-sufficiency low market interaction' by causing zero surplus for sale in markets, land requisition, and flooding have weakened this pattern by forcing peasants to markets. However, local government's embezzlement of compensation had also prevented peasants in Deqing transferring from a high level of self-sufficiency to high interactions with markets. Such situations did not occur in Dongguan. The situation is relatively simple when peasants behave as consumers and Skinner's observations are generally still applicable. The connection between self-sufficiency and market interaction has produced variable effects on rural market advancement in Deqing and Dongguan. Market development has been slow and, as the peasant-markets connection is weak, market hierarchy is rudimentary in Deqing. In contrast, frequent interactions between Dongguan peasants and markets have contributed to the city's rapid market growth and advanced the hierarchy.

'No money, no shopping': Level of income and the marketing pattern of peasants

The level of income of peasants is another major element affecting rural market development (Han, 1998; Zhou, 1998; Zheng, 1997). Skinner does not address this directly because it was not a key factor in a primitive small peasant economy during Imperial China. Instead, he discusses the increase of agricultural commercialisation as a result of a decrease in household self-sufficiency (Skinner, 1965a, p.214). However, greater commercialisation does not necessarily improve the income of peasants. Concomitantly, a rise in income is not naturally contributed to by agricultural commercialisation. This is particularly the case in the reform period, when much of the growth in family income stems from rural industrialisation and remittances from migrant workers.

The income levels of peasants are closely related to their level of consumption. This is regarded by Chinese scholars, such as Gong (1998) and Xiao (1998), as a key factor in the flourishing of rural markets. The higher a peasant's income, the greater is the rate of consumption. A high level of consumption suggests relatively frequent interactions between peasants and markets which is the dynamic for rural market development. Unlike the level of self-sufficiency, this 'high income and high interaction' correlation is applicable to both Deqing and Dongguan. As the income of peasants grow and become more stable, the higher their shopping frequency and the greater their choice of markets. Although the income of peasants in Deqing is generally low, there are remarkable differences between those who have stable incomes and those who do not. Similar situations have also been found in Dongguan, despite peasants being generally wealthier there.

The correlation between the income level of peasants and the degree of market interaction is derived from the marketing pattern of consumer goods in Deqing and Dongguan. Dongguan's relatively higher income results in its peasants spending more in rural markets. In 2000 the income per capita of peasants in Deqing and Dongguan was 3,717 and 6,731 *yuan* respectively (Table 1.1). The consumption expenditure of peasants in the two places accounted for 58 per cent and 75 per cent of their income respectively. In 2000, over 49 per cent of such expenditure was on food while only 6 per cent went on consumer goods such as clothing (SSB, 2001, p.326). Despite the differences in spending on food and clothes, the lower incomes of peasants in Deqing have contributed to lower rates of consumption, implying lower peasant-market interaction. In Dongguan, the situation is reversed. Higher levels of consumption by Dongguan peasants have led to the growth in market activities. As noted in Chapter 4, this factor has resulted in marked distinctions in rural market development between Deqing and Dongguan.

The production patterns of peasants are the prime factors affecting their income and thus their shopping trends. In Deqing, the cash income of peasants remains low as they cultivate paddy fields on a self-sufficiency basis. For example:

> Mr Tan is a typical small peasant living in Xiangshan village of Deqing (Figure 5.1a). While family income was mainly based on sideline production, such as selling cassia bark and keeping pigs, his income was small and fluctuating. Since agricultural production costs and living expenses for his nine family members absorbed a considerable amount of Mr Tan's income, he did not have any extra money to spend at the market.

Thus, the low income of peasants has constrained rural market development because they do not have money to spend. Market activities and turnover are thus stagnant. Conversely, specialised farming and non-agricultural activities have engendered a relatively high cash income for Dongguan peasants. This is illustrated by Mr He's case:

> Mr He ceased farming in 1993. As an entrepreneur (*ge ti hu*) with a delivery business, his income was stable and he was relatively wealthier than ordinary peasants. Living in Hedong village of Wangniudun town, Mr He's monthly income could be over 1,000 *yuan*.

A high income gives peasants more spending power in markets and thus contributes to Dongguan's flourishing rural markets.

Just as production patterns affect the income of peasants, cash income, in turn, affects rural market development. Recalling Mr Tan's example:

> Mr Tan did not visit rural markets regularly because he had no money to spend, particularly for buying consumer goods like clothes and footwear. According to Mr Tan, he bought new clothes and footwear only when the old ones were unfit or worn out. Last year, Mr Tan purchased only one new piece of clothing in the rural market during the Chinese New Year. Family income also determined Mr Tan's choice of markets. As Mr Tan earned only a pittance, he always visited Guanxu market (i.e. the local venue) for everything he needed (Figure 5.1a). Mr Tan never travelled there by bus, although Xiangshan village and the market are linked by a rural road and a private bus service is available. Cycling 9 km to Guanxu market took Mr Tan about 40 minutes but this practice enabled him to save the 4 *yuan* bus fare for other purposes.

Low peasant-market interactions suggest a low consumption level in rural markets. This is the major factor for Deqing's stagnant market activities and development.

Deqing's 'low income and low market interaction' correlation and its impact on rural market development is further illustrated by Mr Liang's case.

> While Mr Liang's family income was based mainly on pine resin production – a popular sideline in Deqing – it was tiny and irregular. The reduction in the purchasing price for resin had further decreased his income from this source to only 170 *yuan* in 1997. Although his two sons were working as migrant workers, no regular remittance was received. His unstable family income constrained Mr Liang's choice of markets and number of shopping visits. Despite Guanxu market being located 5 km to the east of Baxiandong village, it is the nearest for Mr Liang and he used it for everything he needed (Figure 5.1a). Although Mr Liang knew there was a greater variety of clothes and footwear in Decheng, he rarely went there because it is situated 18 km from his home. Also, Mr Liang never travelled to the market by bus because he did not want to spend extra money on transport. He preferred to retain it for other items, as his cash was limited. Family income also constrained Mr Liang's shopping frequency. He usually visited markets every ten days. He did not remember how many times he purchased consumer goods in the market but said it would not be more than three times each year.

This situation has contributed to the persistence of markets in Deqing as places for simple surplus exchange.

The 'low income and low market interaction' relation is also applicable to low-income peasants in Dongguan, despite them being wealthier than their Deqing counterparts.

> Mr Jiang of Hexi village is regarded as a low-income peasant in Wangniudun town. He is a paddy farmer. With less than one *mu* of paddy field, Mr Jiang's annual yields were not very high. As Hexi village – the collective unit – paid peasants' taxes and local levies, Mr Jiang kept all the rice he produced for his own use and for sale. During slack seasons, Mr Jiang also took part-time jobs to earn extra income. Mr Jiang said that the choice of market and shopping frequency depended on his income. Using a bicycle as the major means of transport, Mr Jiang usually purchased foodstuffs, clothes and other daily necessities in Hexi market, which is located about 2 km from his village (Figure 5.2). He seldom visited other venues because he did not want to pay the bus fare. The daily market in Dongguan enabled Mr Jiang to purchase meats and vegetables everyday. However, owing to his low income, he seldom bought consumer goods such as clothes, footwear and household products. Last year, he purchased only a new jacket and a new pair of shoes on the eve of the Chinese New Year.

Mr Jiang is representative of a small number of low-income peasants in Dongguan. However, when compared to his Deqing counterparts, Mr Jiang still had frequent interactions with rural markets despite his relatively low income. Such high interactions between peasants and rural markets are a precondition for market growth. Indeed, Dongguan's 'high income and high market interaction' among its peasants has boosted market activities. For example:

> Unlike peasants in Deqing, Mr He of Hedong village in Wangniudun travelled to different markets for different goods (Figure 5.2). Mr He had rebuilt his home as a concrete structure of three storeys a few years ago. Some of his electrical appliances, such as colour television, were bought from Panyu city, a prefecture-level city located at least 200 km to the west of Hedong village. According to him, there was a large electrical appliances market in Panyu and prices were cheaper than locally. Although transport costs and toll payments were involved, he did not mind travelling a long distance and spent 10,000 *yuan* for his favoured big screen Japanese television. Sometimes he travelled 10 km to Guancheng, the city-government seat of Dongguan, for other electrical appliances. Mr He said his case was not exceptional because many villagers in Hedong village did the same.

Such a shopping pattern has undoubtedly contributed to Dongguan's prosperous market activities.

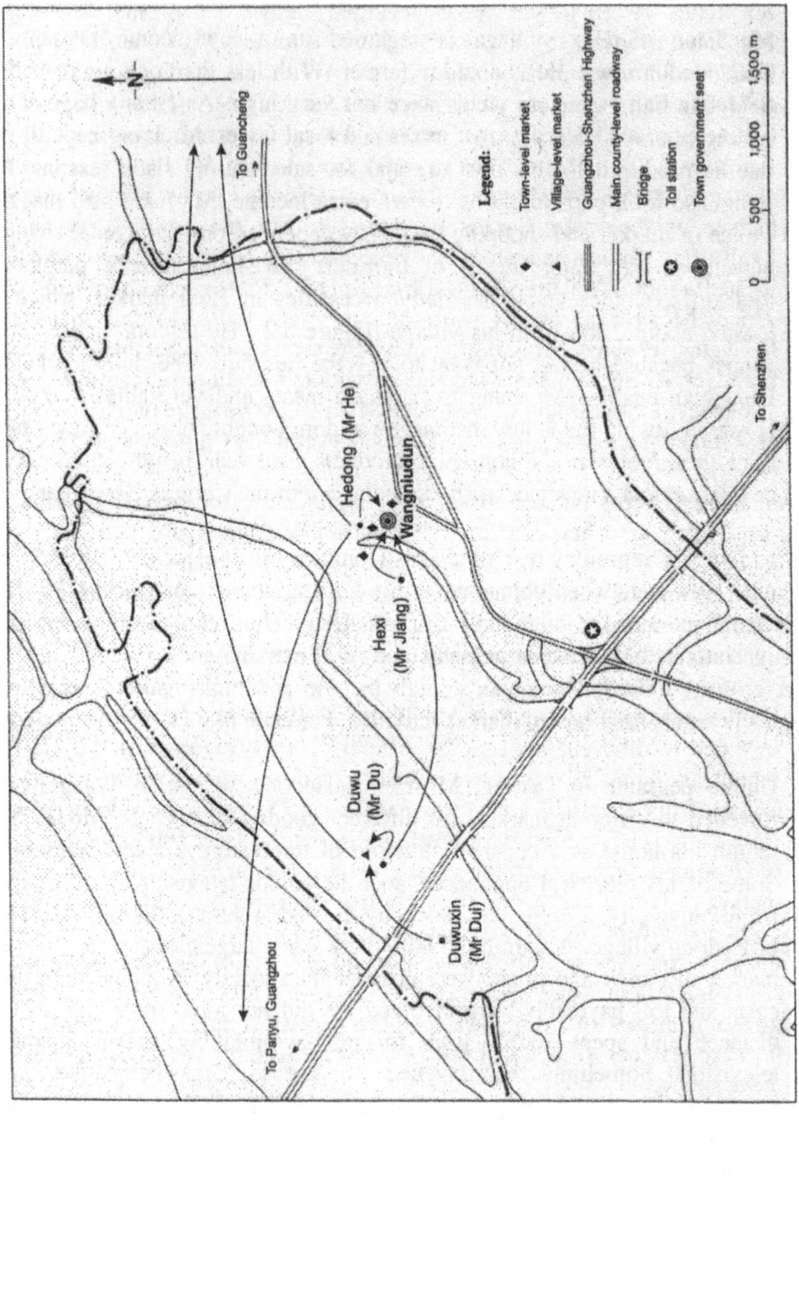

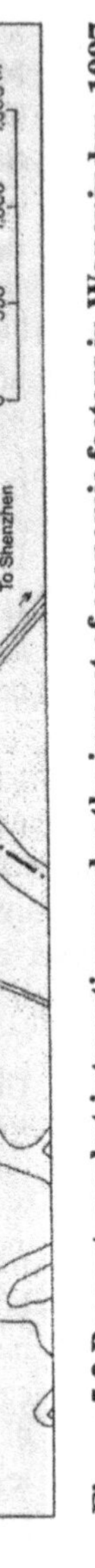

Figure 5.2 Peasant-market interactions under the impact of economic factors in Wangniudun, 1997

The small number of peasants who enjoy a relatively high or stable income exacerbates the slowness of rural market development in Deqing. Unlike Dongguan, there is no 'big farmer' in Deqing. Families who have members working as migrant workers are generally better off. The following cases illustrate this.

> Being a migrant worker for several years, Mr Qi had saved enough money to buy a used motorcycle and start up a passenger delivery business. The income from this is relatively stable so he could afford to visit different markets. Mr Qi usually purchased non-staple foods in Guanxu market and consumer goods in Decheng central market. His shopping frequency was markedly different from other peasants as he also owned a motorcycle. Not only did he visit Guanxu market almost everyday, he also went to Decheng markets four to five times per week, despite it being located 9 km from Tongyougeng village (Figure 5.1a). Transport was not Mr Qi's major concern, he chose Decheng Central Market for consumer goods because there was greater variety and prices were generally cheaper than in the local market. Still, Mr Qi's average expenditure of 10 *yuan* was too small to push market growth.
>
> Remittances from Mr Li's son, who was working in Guangzhou as a truck driver, have enabled Mr Li to travel to different markets by bus. Although the distance between Shangxie village and Decheng market is 16 km and the return bus fare was 6 *yuan*, Mr Li did not hesitate to visit the market and purchase clothes for his grandson. According to him, product quality was better in Decheng markets and he did not mind paying a little more. Foodstuffs were purchased in Guanxu market which is located 3 km on the west of Shangxie village (Figure 5.1a). Mr Li generally visited rural markets during market days. The different timetables in Guanxu and Decheng enabled him to visit both venues without any time conflict.

Their relatively stable incomes have enabled both Mr Qi and Mr Li to have frequent interactions with rural markets and thus spend more money there. Such connections are essential in triggering market growth in a semi-closed economy like Deqing.

Mr Jiang of Zhujiang village also visited different rural markets for the goods he needed.

> He too enjoyed a relatively stable family income as his son was working in Decheng town. This enabled him to travel to other rural markets frequently. He usually shopped in Guanxu rural market but sometimes he visited Decheng markets. Guanxu and Deqing markets are located 5 km and 18 km from Zhujiang village respectively (Figure 5.1a). Although Mr Jiang cycled

> to Guanxu, he did not hesitate to take buses when travelling to Decheng. Mr Jiang said that he enjoyed shopping there because the market is bigger than in Guanxu and there is more choice. He visited the outlet at least once per month. Conversely, he visited the local market not more than three times per week, purchasing non-staple foods, such as pork, for family consumption.

The frequent interactions of peasants with markets suggest they spend more money there. These interactions stimulate the growth of rural markets because commodity exchange is encouraged.

The above discussions demonstrate how the level of income in Deqing and Dongguan affect the shopping frequency of peasants and their choice of markets. This is particularly the case when they are acting as consumers. People with a stable income visit markets more frequently; those who earn a pittance do not have a regular shopping schedule. Also, the wealthy visit different markets for specific items while those with low-incomes are confined to local markets for their everyday needs. Such shopping patterns have different impacts on rural market development in both Deqing and Dongguan. All of these case studies demonstrate that the level of income has become a more significant factor in China's rural economy since economic reform.

Transport : A real or imaginary factor

Earlier criticism of Skinner's overemphasis on transport factors in rural market intensification and modernisation does not mean that they are insignificant. In fact, the significance of transport improvement on shaping economic activities has long been recognised by scholars (Owen, 1968; Fulton, 1969; Bromley, 1971; Hajj and Pendakur, 2000). Skinner also suggests that there is a positive correlation between roads and rural market advancement. In general, he argues that new transport connections stimulate rural market growth such as market periodicity, intensification and modernisation (Skinner, 1964, p.11; 1965a, pp.213, 216, 219). In particular, standard markets wither away when new transport connections to higher level markets are established (Skinner, 1965a, p.220). Such arguments form the core of his analysis on the processes of market evolution and modernisation.

Skinner's transport-market equation is inadequate because fundamental changes have occurred in basic conditions since he established his position. Transport factors include road, accessibility, bus service, and transport modes. Skinner concentrated on road and accessibility but did not anticipate public transport. During the reform era, transport improvement and its economic impact on the Pearl River Delta were identified by Loo (1999) and Lin (1997c) respectively. However, the case of Deqing shows that the positive relationship between transport development and economic growth, as suggested by Lin (1997c), does not necessarily exist in rural market development. In fact, the opening of the New Road 321 appears to have had little effect on local market evolution, though it provides connections to higher level markets. In some remote villages, however, Skinner and Lin's simple equations are still applicable to interpreting the marketing pattern of peasants.

The former situation is illustrated by the case of Jiushi town. While the Huilong section of the New Road 321 stemmed from the upgrading of an existing one, the Jiushi section was new. Before the opening of the new road, Jiushi town was regarded as being in the county's 'corner' – a situation similar to the traditional period described by Skinner. There was only one rural road, and external transport was dependent on waterways before the new one was constructed. Dùwu, Lunchong, Yalizui, Niutoutang and Xiyan villages are located along this new roadway, yet, despite their perfect location, peasants in these places thought the new roadway had had little or no effect on their marketing patterns.

> According to Mr Zhu of Yalizui village, the opening of the new road has neither increased his choice of market nor shopping frequency, despite significant improvements being made to the local transport system. Before the opening of the new road, water transport dominated and 'market vessels' between Jiushi town and Decheng town were available. The journey took about 3 hours and there was only one trip during market days. Road transport has dominated since the opening of the new road. A bus service has been introduced and provided a fast way to visit Decheng town – the county seat that is located 20 km from Jiushi town. Despite buses stopping in front of Mr Zhu's home, he rarely travelled by bus because he could not afford the fare. Accordingly, all the changes appear to have had little effect on Mr Zhu. He continues to visit Jiushi market twice weekly by bicycle (Figure 5.1b).

This demonstrates that rural markets do not necessarily receive a development boost as a result of transport improvements because peasants' interactions with other markets, particularly those at higher levels, may not have risen.

> Mr Lin of Niutoutang village offered a similar observation. He said passing vehicles had increased substantially after the opening of the New Road 321. Also, the number of people who stopped off in Jiushi town had increased. Despite these changes, the new roadway had had little effect on rural market development because passing people are not the major participants in the market. The opening of the new roadway had not altered Mr Lin's shopping pattern. He still kept on using his old bicycle and went to the Jiushi market for everything he needed (Figure 5.1b). Like Mr Zhu, Mr Li rarely travelled to other markets because he could not afford the fare (for instance, a return bus fare to Decheng town was 10 *yuan*).

The consistent use of primitive transport – bicycles – and low incomes are the major reasons why road improvements have had little effect on rural market development. Although an impressive roadway has been built and bus services have been introduced, very few peasants have benefited from these changes. Every day when classes in Jiushi's junior high school were over at four o'clock, the New Road 321 was full of bicycles. To many Deqing peasants, there is no difference between cycling on unpaved rural roads and concrete highways. In Dongguan though, where people have relatively higher incomes and their own motorcycles, the impact of modern roads and bus services on rural market development is substantial. This is because the new roads have increased peasants' choice and degree of interaction with markets. For example, bus services in Wangniudun town have not only linked the town-government seat with villages under its administration, but also connected the town with Guancheng, the city-government seat. The distance between Wangniudun and Guancheng is 13 km and the return bus fare 5 *yuan*. Buses left Wangniudun every two hours and were welcomed by peasants.

> According to Mr Du of Duwu village, he seldom visited Guancheng before the introduction of bus services. As these now provide a convenient and fast way to travel to Guancheng, he visited the city at least once a month (Figure 5.2). He said market-places in Guancheng had a greater and different variety of commodities.

Variations in Jiushi and Wangniudun towns demonstrate that modern roads and bus services have impacted differently on the choice of markets by peasants, and on their shopping patterns. In Jiushi, the insignificance of transport factors is indeed the result of a mismatch between transport development and socio-economic needs of rural communities. While Deqing's peasants still received pittances from sideline production, bus fares were beyond their budget. As well, a modern highway means little to many peasants because bicycles are their major transport mode. This situation shows what Owen (1959) regarded as an ineffective transport system and is common in the Chinese countryside (Hajj and Pendakur, 2000).[4] Accordingly, no significant changes are observed in Deqing's rural market development.

There are other transport improvements, such as the introduction of bus services, which have had little impact on Deqing's rural market development. As noted, the continued use of bicycles by peasants and their low incomes are responsible for this phenomenon. This is particularly the case in the hilly areas of

---

[4] Owen (1959, p.186) argued that "the provision of good transportation is not capable by itself of promoting economic growth". "How effectively the transport system functions" is a critical factor.

Deqing. For example, private bus services have been introduced from the hill town of Huilong but not many peasants use these. In fact, they still visit small rural grocery shops which sell daily necessities such as stationery, shoes, metal fittings, light-bulbs, candles, cigarettes, fireworks, wines, candy, beer, sugar, salt, cooking oil and soy sauce. According to Huilong ICMB officials, there are about 100 small rural shops scattered throughout the town. All of them are run by peasants on a part-time basis.

> Mr Tan was living in Chencun village. The distance between his home and the nearest rural market is over 15 km (Figure 5.3). Like many peasants in the area, Mr Tan purchased non-staple foods and other household articles in the rural shops in Bin administrative village, which is situated 2 km from Chencun village. The recent introduction of 'small wheels' – a private bus service – had almost no impact on Mr Tan's shopping patterns. Nor did he travel by bus to visit different rural markets. According to Mr Tan, the 10 *yuan* return bus fare was a huge amount of money to him as he earned a pittance from his sideline production. Mr Tan's case is typical since many Huilong peasants are small farmers whose income is meagre. Thus, the bus services are of little use because they cannot afford the fare.

Expectations of accelerating rural market development have not eventuated.

> Mr Lu owned a small rural shop of 5 m^2 at the roadside of Shengtian, his home village (Figure 5.3). Everyday at noon, when classes in Shengping's only primary school were over, his shop was crowded with children, buying candles and snacks. His grocery not only sold simple daily necessities, but also was the sole place for social gatherings. According to Mr Lu, although all commodities in his shop were purchased in Decheng markets in the county seat, he did not visit them regularly. This was because turnover was low in the rural area. Although 'small wheels' to Decheng markets were available, Mr Lu seldom took the bus because, like many others in the area, he did not want to spend the 10 *yuan* bus fare. Cycling to the market took over an hour but Mr Lu did not think it was a problem as all commodities he sold were small items.

Skinner's transport-market equation does not mention the decline of roads and its impact on market evolution. Based on his positive correlation between road routes and market evolution, it is anticipated that road deterioration constrains the process of market intensification. However, the decline of roads has not necessarily impacted negatively on rural market intensification in Deqing. When New Road

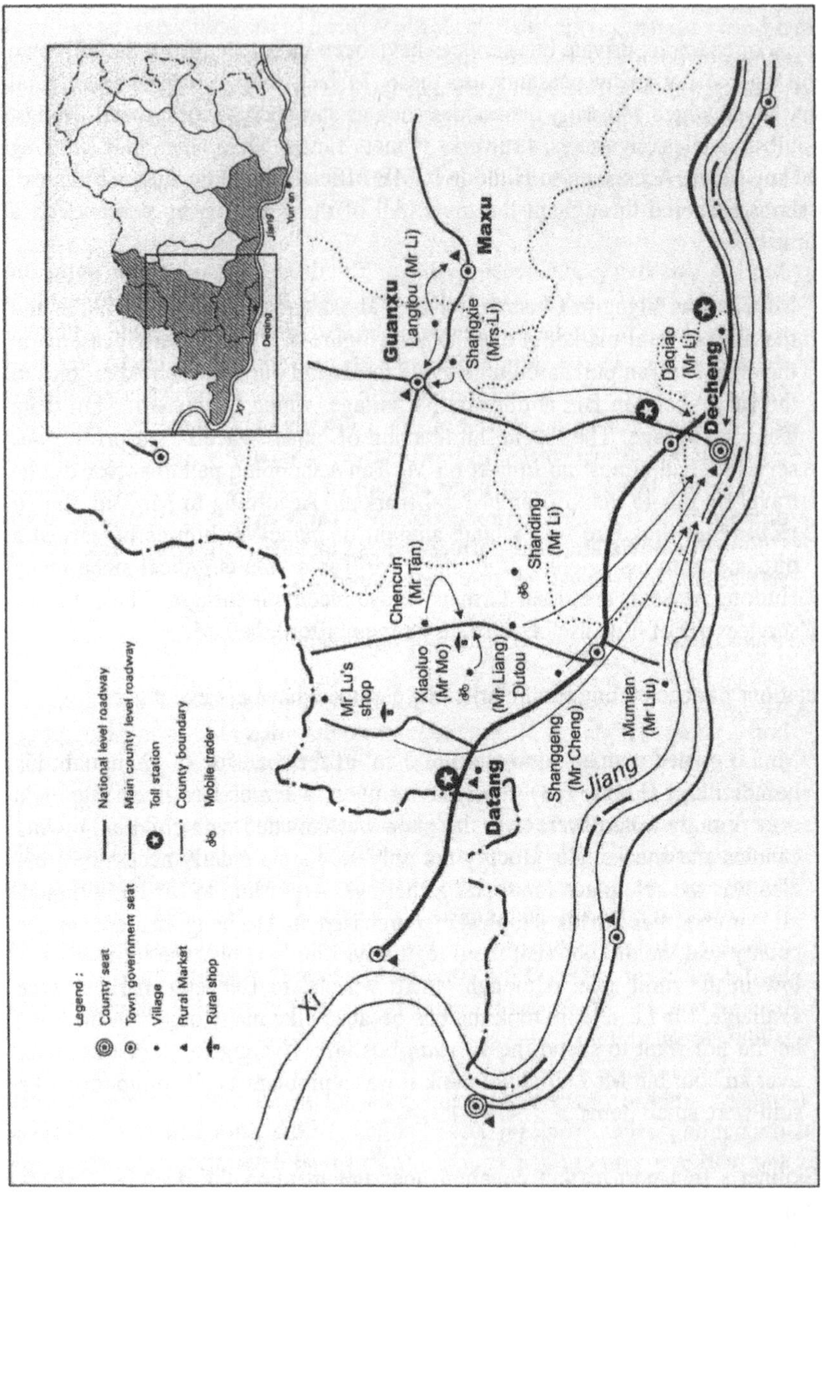

Figure 5.3 Peasant-market interactions under Deqing's simple transport network, 1997

321 was opened in 1995, the old National Roadway No.321 (hereafter Old Road 321) was no longer the sole roadway linking Deqing and the other counties. All vehicles are now directed to New Road 321 which runs alongside the Xijiang River. Old Road 321, which runs over hills inland, has been downgraded to a county-level roadway and traffic has decreased significantly. Surprisingly, the diversion of traffic has had little impact on the shopping patterns of peasants. Thus, Skinner's argument that rural markets decline when they are bypassed by modern roads, has yet to be verified in Deqing.

The reason for this situation is that Guanxu market was established by administrative principles rather than economic forces. Peasants seldom bypass it because it is the only one in the town, the nearest alternative – Maxu rural market – is located some 8 km to the east. Discussions with peasants in Guanxu suggest that the immediate effect of the opening of New Road 321 was a reduction of passing vehicles and people, as a result of which, a number of restaurants and grocery shops in the town-government seat were closed. Yet, Guanxu rural market is still one of the busiest in Deqing regardless of the changes in the road system. In 1997, its turnover value was over 127 million *yuan*, ranking it fourth among the county's 17 markets.

The following cases are good examples:

> Mrs Li lives in Shangxie village, which is located 3 km to the east of Guanxu market. The reduction of traffic on Old Road 321 had no effect on Mrs Li's choice of market and shopping frequency. She still shopped at Guanxu market for sugar, salt and pork, and sometimes clothes, because it was the nearest (Figure 5.2). Also Mrs Li did not alter her shopping frequency because family income was the determining factor. Accordingly, Mrs Li continued to visit markets irregularly.
>
> Mr Li of Langtou village lives 4 km to the east of the market. Since his house is situated less than 100 m from Old Road 321, he was clearly aware of the reduction in traffic. According to him, this only affected some shops in the town but not his shopping pattern. He continued to visit Guanxu rural market once or twice a week on his old bicycle (Figure 5.3).

These illustrations suggest there has been little diversion of traffic following the opening of New Road 321.

Unlike Deqing, however, transport improvements have had a positive impact on rural market development in Dongguan. Wangniudun town is a case in point.

> Mr Du is a specialist banana grower. He cultivated 1 *mu* of farmland and grew more than 200 banana trees. Living in Duwu village, Mr Du was well connected to both local and external markets by paved rural roads which were improved during the mid 1980s. These improvements had enabled Mr

> Du to use big trucks and to engage in long-distance trading. Wholesalers within and outside Dongguan were attracted to the village and purchased products from peasants like Mr Du directly. A purchase point was established in the village and Mr Du used this to trade with them. Prices were negotiated between the peasants and the traders. In 1998, Mr Du sold over half of his 6,000 *jin* yield to wholesalers. As a specialised farmer, Mr Du did not know the final destination of his bananas. He knew only that some were despatched to the eastern and northern parts of Guangdong province. As wholesalers came with big trucks during the banana harvest, peasants like Mr Du did not necessarily concern themselves with delivery and transport costs.

According to Wangniudun ICMB, almost all villages in the town were accessible by paved roads and trucks, allowing peasants to market their products more widely.

Apart from selling to wholesalers, Mr Du also trade a considerable amount in Wangniudun rural markets (Figure 5.2).

> According to Mr Du, although sales were not as large as in local rural markets, the price was higher than that offered by wholesalers. Unlike his Deqing counterparts, Mr Du said transport no longer constrained the marketing of his products because over 70 per cent of the peasants in Duwu village owned motorcycles. The 3 km distance between Duwu village and the town-government seat of Wangniudun took Mr Du only seven minutes. The convenience of such transport had encouraged him to sell some of his bananas in the local rural market for higher profits.

Thus, the better road system and modern transport have enabled Mr Du to sell his products at different markets – some involve long-distance travel and other short-distance. These factors have stimulated the rapid growth of rural markets in Dongguan during the reform era.

The positive relationship between transport and rural market development suggested by Skinner is also valid when peasants are acting as producers. In Deqing, many producers have recognised the significance of roads in market development because poor road systems and lower accessibility have limited their choice of markets. This lack of choice has contributed to low interactions between peasants and rural markets and subsequently impeded the latter's development. For example:

> Mr Mo and Mr Li are small peasants living in Xiaoluo and Shanding villages (Figure 5.3). The distances between these villages and Decheng markets are 23 km and 22 km respectively. Long-distance travel and poor road conditions have discouraged Mr Mo and Mr Li from selling their pigs – their only sideline product – in rural markets. Although 'small wheels' to Decheng market were available, Mr Mo never considered this option

> because the return bus fare was 6 *yuan*. As a producer, Mr Mo said he had to minimise costs. Neither did he cycle to the Decheng markets because it took almost two hours. Mr Li of Shanding village also said the rough and narrow road surface had created great difficulties for him in delivering pigs by bicycle. Consequently, Mr Mo and Mr Li sold their animals to mobile traders, despite the price being slightly lower than at the market.

According to Huilong ICMB, there were about 10 registered mobile pig traders in the town. They travelled around the hilly villages by motorcycle, buying pigs and selling pork to peasants. Since interactions between producers and markets have been limited, market activities have stagnated. Consequently, market growth has been very slow.

Conversely, where peasant-market interactions are relatively intense in villages with higher accessibility and served by better rural roads, market activities have flourished because pigs are sold locally. These interactions provided the dynamics for rural market growth. Despite mentioning the importance of transport factors in their decision making process, peasants often have other concerns. This is illustrated in Mr Liang's case:

> Mr Liang lives in Dutou village of Huilong town. Despite the distance between Dutou village and Decheng market being about 17 km, the village is linked to the New Road 321 by a better rural road, and 'small wheels' are available. Superior road connections had attracted Mr Liang to sell his two pigs in Decheng market last year (Figure 5.3). Although he said transport factors were undoubtedly important, as a producer he had other concerns. Also, he chose Decheng market because it is the county's largest rural market – the trading scale is bigger than in other town-level markets and the selling price is much higher.

Mr Cheng of Shanggeng village had a similar point of view:

> Living 13 km to the west of Decheng market, Mr Cheng sold four pigs in the market last year (Figure 5.3), which were delivered to the market by bicycle. Since Mr Cheng's home is situated along New Road 321, he said it was easier to cycle on the concrete road surface, though it took him 30 minutes to reach Decheng market. Mr Cheng also thought transport was not the sole reason for his choice. The other factor was the concentration of non-agricultural population in Decheng town, the county seat. As Decheng town is a big outlet for agricultural products, the turnover rate is higher than in other markets. Last year, Mr Cheng made about 200 *yuan* profit from selling his four pigs.

The frequent interactions of producers with Decheng markets have contributed to its reputation as the busiest in Deqing; it is also the largest among the 17 outlets in the county.

The effect of variations in transport routes on rural market development is further illustrated by the case of a specialised farmer.

> Mr Li, who lived in Daqiao village, was one of the very few specialised farmers in Deqing county. The village is located on the county seat's eastern border and links Decheng markets by New Road 321 and other paved rural roads. Also, public and private bus services are available. The concentration of non-agricultural population in Decheng town had given rise to a greater demand for vegetables and this prompted Mr Li to shift from mono-cultivation of paddy to specialised farming several years ago. Moreover, the extension of the Decheng joint-dyke to Daqiao village has resolved the flooding problems that the village had suffered for decades. A lot of households in Daqiao village have changed therefore to specialised farming as both transport and production conditions have improved. Mr Li had subcontracted 1 *mu* field for vegetable growing. According to Mr Li, he grew different vegetables in different seasons. This practice enabled him to have a variety available in markets throughout the year. All produce was sold in Decheng central market (Figure 5.3). In 1997, Mr Li sold over 6,000 *jin* of vegetables and some of his neighbours sold over 10,000 *jin* in markets.
>
> Mr Li was delighted to see the transport improvements in recent years and thought such changes had enabled the efficient delivery of his vegetables to the market. As he rented a fixed stall in Decheng central market, he and his family members had to deliver perishable vegetables early every morning. Mr Li insisted on delivering these by bicycle, although bus services were available in Daqiao village. Sometimes his brother helped to shoulder-pole the vegetables to the market. As the distance between Daqiao village and Decheng central market is only 2.5 km, cycling and walking took 15 and 30 minutes respectively. Given the semi-closed conditions of Deqing county, better rural roads and transport connections have become the major factors in Mr Li's decision to sell his products in Decheng market.

Market development is certainly boosted when specialised farmers are connected to the rural market by a better road system and transport services.

Mr Liu's case is another example:

> Mr Liu was one of the very few specialised farmers in Huilong town. Unlike his Dongguan counterpart, Mr Liu still grew grains and vegetables for his own consumption. In 1995, Mr Liu borrowed 90,000 *yuan* from the local credit union and subcontracted 70 *mu* of farmland for fruit growing. About 50 *mu* of sloping land was devoted to lychees and 20 *mu* of flat land to

vegetables and a variety of fruits such as pear and pomelo. He also used a small proportion of the loan to pave the small road in front of his house and enhanced his irrigation system. Mr Liu and the families of his two brothers, who live in the same village, formed a working team of 15 to look after the huge farm. He hoped that the 70 *mu* farmland would earn him a fortune within 20 years.

Living in Mumian, Mr Liu said the village was in a perfect location. Huilong's old town-government seat was situated about 500 m on the east of the village. As the old town-government seat was situated along the Xijiang River, both road and water transport were available. Before the removal of the old town-government seat, there was a concentration of government offices and small shops. As noted, geographical proximity to the town-government seat provided markets for Mr Liu's products. Although the market was not as big as that in Decheng, it also attracted peasants from nearby villages. Shops, and booths selling meat and vegetable were also available. However, in 1995, the town-government seat was moved to a new site located along the New Road 321 and 3.5 km from the original. In a small town like Huilong, the relocation of the town seat ruined all the economic activities at the old spot, peasants no longer went there and all shops were closed. At the new site, another market has not yet been formed, though all government institutions have been established. Only two small groceries and one small restaurant exist there and these are not usually patronised by people. Interviews with peasants suggested that they had switched to either Datang or Decheng markets situated 6 and 12 km respectively from the new government seat. Such changes provide a good illustration of the impact of government intervention on rural market development, suggesting that market demise results from government intervention rather than from the transport improvements highlighted by Skinner.

Poor roads and transport facilities in Huilong town had created extra difficulties for Mr Liu in delivering his products to rural markets. This situation had constrained market development because interactions between peasants and rural markets were limited. Despite Mumian village being located along the Xijiang River and accessible by both road and water routes, neither of them were well developed. Vessels to Decheng markets were only available on market days. Although the delivery of goods by water was cheaper than by road, the vessels were small in size and slow. For example, both water and road distances between Mumian village and Decheng town are about 10 km. Road transport took only 25 minutes but water transport took almost an hour. Last year, Mr Liu delivered some vegetables and pears to Decheng central market by water (Figure 5.3). According to him, the longer travelling time by the water route made him miss the peak business hour in the market. Since Mr Liu's products are perishable and Decheng is a large competitive market, his late arrival meant a loss of business and profits.

Road transport is also poorly developed in Mumian village. The opening of New Road 321 (Huilong section) has not increased the accessibility of Huilong town and Mumian village.

> As the village is located 3 km away from the roadway, Mr Liu had to use the rural road before reaching New Road 321. According to Mr Liu, he sold 3,000 *jin* of jujube to Wuzhou city last year – this was delivered by truck. However, the vehicle was too big to reach Mumian village because the rural road link to New Road 321 is narrow and rough. Mr Liu had to use bicycles and tractors to take his products to the rural road before they could be transferred to a truck and delivered to their destination. This practice cost Mr Liu more than 700 *yuan*. The situation is the same for other products delivered to Decheng market by road. According to Mr Liu, this dilemma will persist until there is a significant improvement in Huilong's road system and in Mumian village's accessibility.

While transport problems create obstructions for specialised farmers delivering their products, the growth of both specialised farming and rural markets were constrained. Rural markets cannot expand when the levels of specialisation and commercialisation are so feebly developed. A similar conclusion was drawn in Lado's (1988) study of rural market development in southern Sudan of Africa.

On top of the zero impact of transport improvement on Deqing's rural market development, it is worth mentioning that remoteness and poor transport facilities have led to agricultural specialisation in the hilly areas. However, the level of specialisation has not been as good as in the Delta. Indeed, the service area is concentrated on only a small number of off-road villages. Specialisation is limited to pork, and mobile traders travel around different villages in the hilly areas by motorbike, buying pigs and selling pork to peasants. A small table is installed on the back seat of the bike and baskets are added to both sides. According to officials at Huilong industrial and commercial office, no statistics are available on these mobile traders because they do not need a special trading licence. Some of them, however, are illegal because they have not registered with the appropriate authorities. Remote villages within Huilong and Guanxu towns are all served by mobile pork traders who are welcomed by the peasants. The emergence of mobile traders in areas of poor accessibility was not anticipated in Skinner's original framework. Neither was it observed in the rural marketing network of other developing countries such as Nigeria (Porter, 1995). Whether this primitive specialisation will eventually progress to a formal marketing activity is hard to predict. However, remoteness and poor transport facilities in Huilong contribute to the current poor market development in these hilly areas.

In short, the impact of transport factors on rural market development in this study has been examined by considering the choice of markets by peasants. The significance of transport factors varies between the study areas. In Deqing, modern roads and transport modes have had little effect on boosting market activities because interactions between peasants and rural markets have not intensified through these changes. Yet, the same factors play a more important role in Dongguan because the level of development and incomes of peasants are different from those in Deqing. The impact of transport factors also varies according to whether peasants are acting as producers or consumers. In both Deqing and Dongguan, peasants regard transport issues as important when they attempt to sell their products in rural markets; this is true also of specialised farmers. In contrast, transport factors play a less crucial role when peasants are acting as consumers. Whether this variation is a result of what Lin (1997c, p.162) called the "lag relationship" between transport improvement and economic growth in Guangdong has yet to be proven. Obviously, the significance of transport varies according to different conditions and to the behaviour of the peasants. Given these variations, Skinner's thesis needs careful qualification if it is to reflect accurately the role of transport factors in rural market development.

Skinner's analysis on rural trade and marketing is being criticised as "extremely perceptive" (Hodder, 1993, p.31). While some of these perceptions find support, others do not. Three economic factors, notably the level of self-sufficiency, the level of income and the transport, were addressed in this Chapter. Obviously, the significance of each factor is determined by the locality effects of Deqing and Dongguan. A high level of self-sufficiency constrains Deqing's market development as interactions between peasants and markets are limited by production patterns. This situation is reversed in Dongguan, where agricultural production is highly specialised. Moreover, the level of income of peasants plays an equally important role in affecting their choice of markets and shopping frequency in both Deqing and Dongguan. This income differential creates distinct levels of market development in the two places. Finally, the significance of transport routes, modern road systems and bus services on rural market development varies in Deqing and Dongguan. Generally, such factors have had a more significant impact on rural market development in Dongguan than in Deqing. In the latter, transport improvement projects have created a big gap between transport and socio-economic needs. While peasants are generally in low income and depend on bicycles, these improvements mean little to them. As a result, rural marketing activities benefit nothing from these changes.

The changes in China's social-economic and political conditions during the reform era have created a complicated setting. The cases of Deqing and Dongguan demonstrate that all components play key roles in shaping rural market development. This conclusion has provided strong counter-arguments to Skinner's rather parochial views on the importance of transport elements. Transport was not the only component involved in China's rural market development. Thus, the position of transport factors, as Leinbach and Chia (1989, p.2) and Bryan *et al.* (1997) argue, is "necessary but not sufficient". Moreover, this understanding has weakened the explanatory power of Skinner's theory. This is because the significance of many seminal factors has changed. Moreover, variations between Deqing and Dongguan, and between producers and consumers, demonstrate the complexity of the real world and affirm the oversimplification of Skinner's thesis of rural market development. China's rural market development is not as simple as Skinner's original thesis suggests.

Chapter 6

Invisible Hand versus Invisible Wall: Administrative Parameters and Rural Market Development

This Chapter extends previous discussions by focusing on non-economic elements – administrative parameters and their impact on rural market development (see Chapter 5). Skinner's perception of administrative factors and their impact on rural markets is based on a cyclical theory of policy change. He argues administrative intervention played a crucial role in rural market development during Mao's period which witnessed the domination of an anti-market ideology and the implementation of radical policies such as collectivisation (Skinner, 1985b). However, their significance has been wiped out by the elimination of the above policies and the release of controls on agricultural production and trade during the reform era which was regarded as a liberal phase with "a steady and methodical deradicalisation" (Skinner, 1985b, p.405).

Despite Skinner's meticulous insights, the persistence of government regulation during the reform period (see Chapter 3) does not lend support to his policy cycle theory. In fact, his hypothesis has oversimplified China's political economy by assuming various reform measures have automatically dismantled government control. Although policy changes were considered, the structural forces of the state have been overlooked. These omissions account for Skinner's underestimation of administrative components and their role in rural market development during the reform era.

This chapter demonstrates that administrative aspects – despite being weakened during the reform era – are still crucial in understanding China's socio-economic development. It also exposes the limitations of Skinner's overemphasis on economic and transport factors. The impact of administrative parameters on rural market development is examined through a study of interactions between peasants and rural markets in Deqing and Dongguan. Particular attention is paid to the choice of markets by peasants. Initially, the role of administrative principles on market establishment and development is discussed; this is followed by an analysis of local government regulations.

Administrative principles and the formation of a single market structure

The creation of market places using administrative principles is an important factor that dictates China's rural development. Since Mao's era, the administrative hierarchy has overridden the economic one in China. Such pre-eminence has persisted since the reforms of 1978, resulting in China's rural market system being an appendage of the administrative system. As rural markets are established by regulation under the administrative-zone economy, government rules have replaced economic forces as the dynamo driving market development. These principles have not only determined the number of rural markets in specific towns, but also their location. Unlike markets formed by the 'invisible hand', those set up by government do not naturally decline if they are not patronised by peasants, neither are they relocated or modified to attract patronage. Under this regime of 'one designated town, one rural market', there is usually little choice but to visit the sole market place in the town, even when it is badly situated. These principles apply to all town-level markets throughout China and have had a significant impact on the country's rural market development.

Under the single market structure, people's choice between alternative locations is limited and the number of interactions restricted. This is the main reason for slow rural market development, particularly in Deqing county. Deqing's semi-closed conditions have meant that the county's rural market system is very basic and reliant on the government system. As many towns in Deqing have only one established rural market, peasants have little choice where to buy and sell – a situation further aggravated by their low incomes and reliance on the bicycle for transport.

Peasants using government markets in Deqing have many complaints. Although, as producers, they have little surplus for sale, they would prefer a wider range of alternatives, particularly the opportunity to use larger markets which offer higher prices. The single market structure, however, requires producers to sell at a market in a specific town. There is only limited competition because the commodities being sold are identical and, generally, quality is not high. The massive influx of similar products during particular seasons also reduces the selling price and limits sales. For example:

> Mr Mai, living in Liucunge village of Guanxu town, sold his surplus lychees and mandarins in the local market last year (Figure 6.1a). This was Mr Mai's only choice because the nearest other markets – Maxu and Decheng – are situated 10 km and 15 km respectively from his home. However, he was one of dozens of peasants selling lychees and mandarins. The similarity of commodities in the market had slashed Mr Mai's income because the selling price of lychees had decreased from 1 *yuan* per *jin* to 0.5 *yuan* per *jin*. Mr Li was even less fortunate. Although he was desperate to sell his ginger for extra income, he failed despite offering it for sale at a very low price. He had to return home without selling anything.

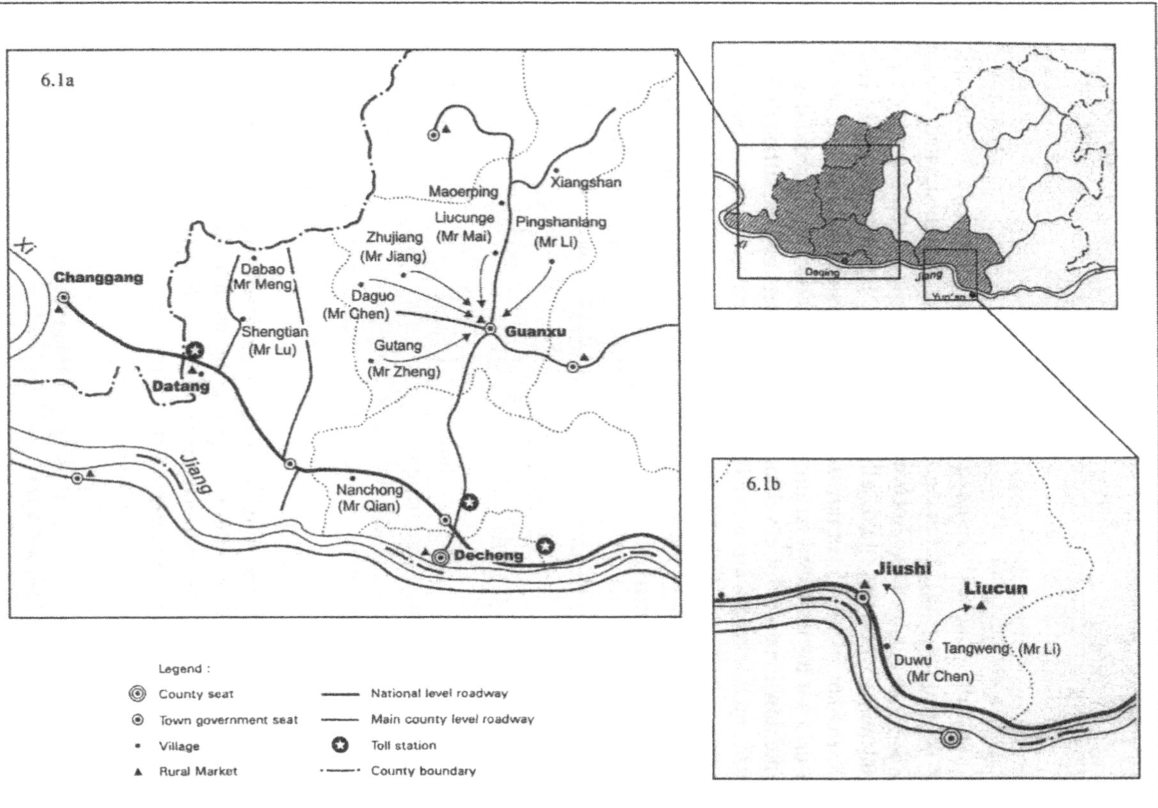

Figure 6.1 Peasant-market interactions in Deqing, 1997

This situation has discouraged peasants from selling their surpluses in rural markets, not to mention the marketing difficulties faced by big farmers and specialised producers. Mr Jiang's case is a good example:

> Despite a bumper harvest for lychees in 1997 with an unusually high surplus, Mr Jiang sold less than half of his surplus produce in Guanxu rural market (Figure 6.1a). Mr Jiang lives in Zhujiang village situated 3 km west of the Guanxu market. According to him, neither distance nor transport determined where he could sell, he had to use Guanxu market as it is the sole market in town. As noted by Mr Jiang, since he was neither a 'big farmer' nor a specialised farmer with many products to sell, he did not want to increase his costs by travelling to other markets by bus, although buses to other town markets are available in the village. Nor did he cycle to the Decheng market because it is situated 15 km south of Zhujiang village. Such settings and a lack of local choice have discouraged Mr Jiang from selling his surplus in the market.

Discussions with Deqing peasants suggest that many of them share his opinion. They thought that the lack of alternative outlets contributed to all of their difficulties associated with selling surplus produce. If producers are not selling their surplus, market development is stunted. Trading – the prior condition of market evolution as regarded by Skinner (1965a) – remains stagnant and improvement in agricultural production is undermined. This explains why market activities in Deqing have not matched those in Dongguan.

Further, markets offering similar local products deter consumers from visiting. For example:

> Mr Li, living in Pingshanlang village of Guanxu since the 1940s, was no longer interested in visiting the town's sole market. As a consumer, he was disappointed with the familiar range of products. According to him, there was no specialisation as peasants from the same town were selling virtually identical products. For consumer goods, such as clothes, the situation was the same. Consequently, Mr Li only visited the local market irregularly to buy non-staple foods, despite earning a relatively stable income from pine resin production (Figure 6.1a).

The mismatch between market location and population distribution also discouraged peasant-market interactions under Deqing's single market structure. As rural markets are always established in the town-government seats, their location offers less than ideal access for the peasant population. This situation is little different from the 1950s when market activities were centralised under the People's Commune. Clearly, the use of administrative measures to control market activities has persisted. Peasant-market interactions are constrained therefore because peasants have to travel long distances to visit the sole market in town. This handicap is particularly pronounced for those living on a town's border. Thus, the optimum market-to-village ratio devised by Skinner (1964) loses its explanatory value for patterns in the reform period.

Take Mr Zheng's case as an illustration:

> Mr Zheng of Gutang village lives 8 km to the west of Guanxu market and he purchased non-staple foods and other daily necessities there because there was no real alternative (Figure 6.1a). He did not consider other markets because they are located in other towns. Using a bicycle as his major transport mode, it took Mr Zheng over 30 minutes to visit the local market. Other town markets are obviously beyond his reach. Mr Zheng's case is a typical example demonstrating the drawbacks of determining market locations by fiat. Peasants in Xiangshan and Maoerping, which are located 7 km and 9 km from Guanxu market respectively, also face a similar situation.

This practice has discouraged peasants from visiting Guanxu market – the only market in the town.

Deqing's single market structure has created a dilemma in the county's rural market development. On the one hand, peasant-market interaction is discouraged by low-level competition due to a mismatch between market location and population distribution. On the other hand, administrative principles also constrain marketing activities in non-designated market sites. Dissatisfaction with the single market structure has triggered the selling of agricultural products at spots outside the official venues. This practice could have a positive effect on Deqing's rural market development as the choice of sites for peddling is economically rather than administratively determined. However, selling at these sites is prohibited by law, a restriction which has jeopardised market activities in non-designated locations because any attempt to establish a 'natural' market system – built on Skinner's economic forces – is precluded. As noted in Chapter 4, this is a prime reason why Deqing consistently has a low market density and high market-to-village ratios.

The imposition of market fees is another reason for people peddling at spots other than rural markets. According to Deqing ICMB, in rural markets peasants are subject to fees – formal payments levied by the Bureau. However, many regard the charges as an administrative measure to control market activities; they do not understand why levies are imposed on non-professional traders like themselves. Interviews with ICMB officials suggest that tolls are an important measure for implementing modern market management. As tidy and well-established places are provided, payments are imposed on those who trade there so that the ICMB has the necessary resources to maintain them.

Market fees include an administrative charge and rent. Generally, they are combined into one single payment imposed on sellers in rural markets. This is supposed to be levied on 1 to 1.5 per cent of a seller's daily turnover value. But

ICMB officials usually levy a 5 to 10 *yuan* impost on peasants instead of checking their actual turnover value. The 5 to 10 *yuan* market fee represent a huge amount to many who trade in Deqing because sales rarely exceed 100 *yuan* per day. Consequently, it is not unusual to hear complaints about the combined fee and, as a result, some peasants sell their surplus illegally at spots outside market-places to avoid this charge.

This is the case with Mr Qian:

> Living in Nanchong village, he is dissatisfied with the single market structure as the nearest rural market is situated 9 km from his village. Instead, Mr Qian has chosen to sell his surplus lychees at the side of the new 321 roadway, the major roadway in Deqing county. Mr Qian did not have a particular selling spot, he usually sat at the roadside and waited for passing vehicles to stop (Figure 6.1a). This practice is irregular and sometimes accidents have occurred. Despite this, Mr Qian sold about 100 *jin* of lychee in this way last year.

Mr Qian's practice is unexceptional. It is common to see peasants selling various kinds of agricultural products on the roadside. This was particularly the case in the summer of 1998, when there were plenty of surplus lychees for sale. However, officials carry out inspections everyday, and if peasants are caught by the ICMB, not only is their produce confiscated, but also a heavy penalty is imposed. Like many who peddle on the streets, Mr Qian had his goods confiscated several times last year. He thought this an overreaction and his relationship with officialdom has become strained. As a result of the crackdown, commodity exchange is highly concentrated at certain venues only, with almost no market activities outside designated market locations in Deqing.

Unlike Deqing, Dongguan's prosperous economy and well-established market system have wiped out the restrictions inherent in the administratively established market system. The emergence of village-level markets reflects the development of a 'natural' system based on economic rationale in Dongguan. This situation has intensified peasant interactions with markets because they have a greater range of choice. Producers, particularly specialised farmers, have a wider range of markets to patronise. As noted in Mr Dui's and Mr Du's cases (Chapter 5), peasants sell their products to markets both inside and outside Dongguan. Given the high level of agricultural specialisation and commercialisation of agricultural activities in Dongguan, it is not surprising that some peasants did not know the final destination of their produce.

Moreover, in Dongguan, the emergence of village-level markets also provided a greater choice of places for small peasants to sell their surplus, although their numbers are few in the city. For instance:

> Mrs Li was selling lychees at Humen long-distance coach station last year. Living in Liangwu village of Humen town, Mrs Li said administratively

> established town-level markets were not the only place to sell her products. The opening of village-level markets provided her with other choices. Sometimes she opted to sell lychees at the roadside market outside coach stations, despite competition being fierce and, reportedly only marginal profits being made.

This case demonstrates that market activities are flourishing both inside and outside designated locations in Dongguan. The intensification of market activities, as acknowledged in Skinner's (1965a) theory of market growth, is the prime reason triggering market evolution.

The impact of Deqing's single market structure on constraining market growth is also illustrated by the government's monopoly of rice selling. While assuming government regulations were dismantled by various reform measures, this factor was not discussed in Skinner's mid-1980s articles. Unlike the marketing of foodstuffs and clothes, rice selling is still controlled by state regulation. Rice is only available in designated supply and marketing co-operatives (*gong-xiao she*) located in town-level markets – the 'official' markets set up by the government. In Deqing, a small number of peasants have to purchase their grain from rural markets because their own production is insufficient, due often to land requisition (see Chapter 5). While rice sales are confined to designated outlets, interactions between peasants and rural markets are restricted to specific places. Such controls have not only 'throttled' the private sale of rice, but have also 'strangled' market activities led by economic forces. For example:

> Mr Chen lives in Dúwu village of Jiushi town, situated 3 km to the east of Jiushi rural market. Despite buying rice in Jiushi market, location and distance were not the prime factors determining his choice of market. Nor had price or product quality governed his decision. According to Mr Chen, he chose Jiushi rural market because it was the only place in town where rice was available (Figure 6.1b). If he bypassed the local market, he had to travel over 20 km to the nearest market at Yuecheng. Using a bicycle as his major transport mode, Mr Chen visited the market once a month. He did not consider other venues because they are too far to reach by bicycle.

As peasant purchasing activity is concentrated on the sole market in town, other alternative sites are not used. This explains why the density of markets remains low in Deqing, and its market system primitive.

Conversely, administrative regulations do not have a significant impact on market development in Dongguan. This is because preferential policies and foreign investment have reinforced the role of the 'invisible hand' in Dongguan's economic development. Although official rice-selling outlets are established in town-level markets, they are not the only places to buy rice. The emergence of village-level markets has provided peasants with more choice than in Deqing; there are always people selling surplus rice there. This suggests that market activities have developed outside the designated market-places. For example:

> Mr Du, living 3 km away from Wangniudun rural market, had two choices of market because rice was also available in Duwu village's rural market. According to Mr Du, he often purchased rice in the village, despite better quality rice being available in the designated town market. The opening of village-level markets in Wangniudun has provided extra convenience to local peasants because they do not need to travel long distances to buy rice. The emergence of village-level outlets also suggests market activities have developed outside designated locations.

The interactions of peasants with these village outlets have encouraged a division of labour and market integration – events which also explain why rural markets in Dongguan are more advanced than in Deqing.

Apart from rice, Dongguan's well-established market system not only provides more choices for agricultural producers to sell their commodities, but also offers consumers a greater range of products to buy. For instance:

> Mr Du of Wangniudun town in Dongguan visited his village-level market for non-staple foods and vegetables. For high-order goods, such as clothing and footwear, Mr Du chose Wangniudun's town-level markets, which are located 3 km from his home (Figure 6.2). Unlike Mr Du, Mr Chen of Daguo village within Deqing purchased all necessary foodstuffs, clothes and other household items in Guanxu market, despite Daguo village being situated 7 km to the west of Guanxu (Figure 6.1a). Although Daguo village is connected to the market-place by a rural road, Mr Chen said this was not his major reason for visiting Guanxu market, he chose it because it is the only place where such products were available.

These examples demonstrate how market settings in Deqing and Dongguan affect the choice of outlet for buying ordinary products.

The use of administrative principles to establish rural markets also discourages market integration and specialisation by eliminating the possibility of peasants choosing alternative outlets in other towns. Again, Skinner neglects such causal relations in his thesis. In Deqing, very few people can afford to travel to markets in different towns, most are small peasants who depend on the local market for everything they need. Although interactions between peasants and rural markets are observed, these are established without choices being available and they do not benefit market growth. Accordingly, differentiation between markets is low and specialisation and integration limited.

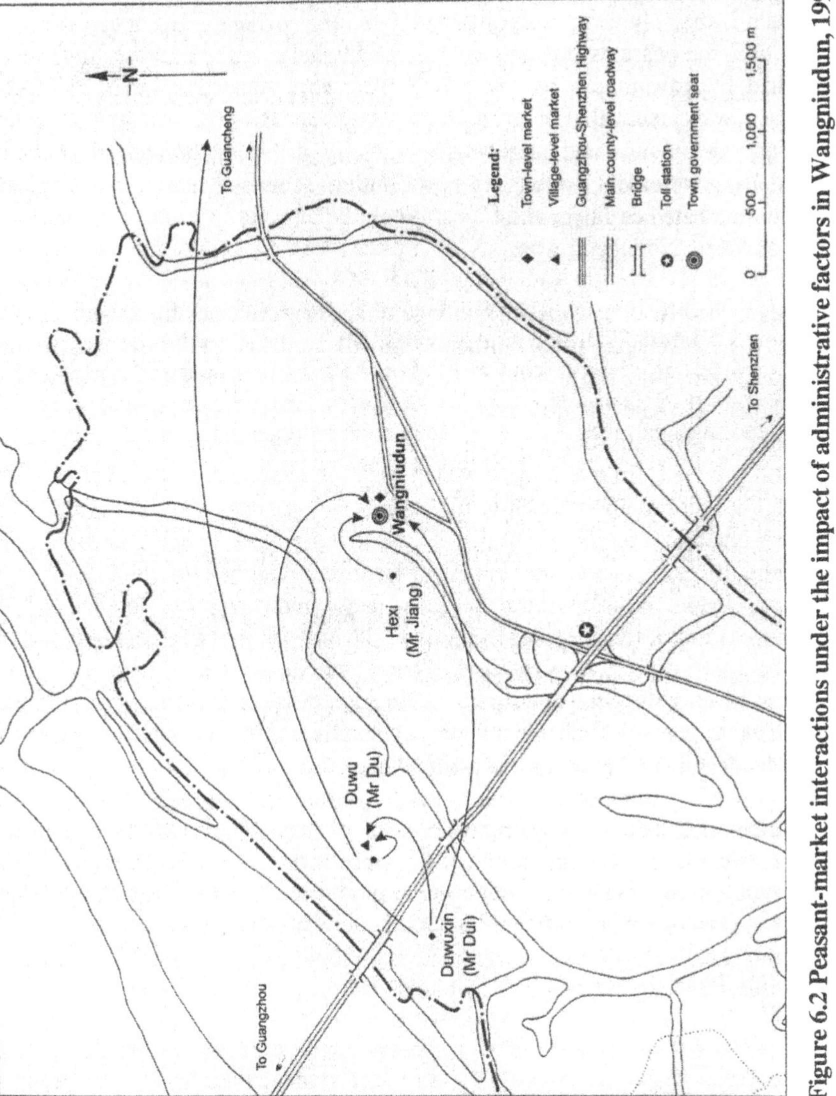

Figure 6.2 Peasant-market interactions under the impact of administrative factors in Wangniudun, 1997

Jiushi is the only town in Deqing with two rural markets – Jiushi and Liucun. Despite this, the peasants do not have any additional choices because both were established by government fiat. Although they were situated 5 km apart, their appearance was identical. For example, farm chemicals were available in both venues and the brands and prices were virtually the same because they were provided by government established retail outlets. This suggests a low level of differentiation between Jiushi and Liucun rural markets, not to mention similarities with those located in other towns.

> Mr Li had lived in Tangweng village of Jiushi town since the establishment of the PRC (Figure 6.1b). Although the village is located 3 km from Liucun and 5 km from Jiushi rural markets, Mr Li did not think he had a greater choice. He said the commodities in both markets were identical and the level of specialisation low. Mr Li purchased farm items, such as fertiliser and pig feed, from Liucun, despite it being regarded as a small market. He did not disclose how much fertiliser he consumes each year, but said it was not much because he had only a 1.6 *mu* low yield paddy field. According to him, the price, quality and brands of fertiliser were the same in both Liucun and Jiushi rural markets because they were supplied by the same governmental retail agency. Mr Li was the only interviewee who still used a shoulder pole to carry fertiliser. This primitive method of delivery meant that he took over one hour to travel the 3 km distance between home and the market. The small differentiation between markets in Jiushi also explains why Mr Li purchased fertiliser from the nearest rural market.

Conversely, in a more commercial area of Dongguan, a higher level of differentiation and specialisation was observed between markets. The implementation of a responsibility system in farm chemicals has resulted in different levels of service, despite the same government-owned farm-chemical supply and retail co-operative being found in all town-level markets. For example, delivery services are provided in Wangniudun town.

> Mr Jiang lives 1 km from Wangniudun's Hexi rural market. He purchases fertiliser there like many peasants in Hexi village (Figure 6.2). Like his Deqing counterparts, neither distance nor price determined Mr Jiang's choice of market because fertiliser is only available there. Cultivating less than 2 *mu* of paddy field, Mr Jiang did not use much fertiliser and he bought it only once a year. Transport and delivery were not a concern because a door-to-door delivery service was provided. According to Mr Jiang, the price of fertiliser was higher as transport costs were included. However, he welcomed the service because he did not have to arrange transport himself. Such a service was not only provided a boon to peasants, but also supplied the dynamics for market growth.

Recently, however, a similar service has emerged in remote areas in Deqing. For example, Dabao village is situated at the hilly border of Huilong town and the nearest rural market is located 8 km from the village (Figure 6.1a).

> Remoteness and poor rural roads had created great difficulties for peasants like Mr Meng in delivering bulky products such as fertiliser. As a result, many peasants bought fertiliser from mobile traders who usually visited the village at the beginning of the cultivation season with a full tractor load. Retail prices were slightly higher than those in rural markets and supply co-operatives because transport costs were included. Cultivating less than 1 *mu* of field, Mr Meng did not really care about the price difference because he used less than 700 *jin* farm chemicals each year. Since arranging transport is a problem, Mr Meng said he had no other choice, whatever the price. He chose to buy from mobile traders because roads were poor and transport difficult.
>
> Mr Lu of Shengtian village also chose to buy from mobile traders. Since he had more than 4 *mu* of paddy field, he used in excess of 1,000 *jin* of chemical fertiliser each year. Although this was not a huge amount, poor rural roads and transport facilities meant delivery was very difficult. Mr Lu suggested mobile dealers charged 5 to 6 *yuan* per bag more than those in the rural market. However, to avoid transport problems, many peasants in Shengtian and other villages still bought from mobile traders.

Whether such a development will eventually alter Deqing's market development is yet to be seen. Given the semi-closed nature of Deqing's economy and the low consumption level of peasants, many adjustments still have to be made to ensure market growth.

The above discussion demonstrates the impact of administrative principles on rural market development in Deqing and Dongguan. As administrative principles determine the number of rural markets and their locations, these factors also decide where peasants go to buy and sell products. Although government regulations established stereotypical markets at the town-level and provided the same basic infrastructure in different places, the study of peasants in Deqing and Dongguan showed that their impact varied from place to place. The highly commercial environment of Dongguan has overcome many of the drawbacks and restrictions associated with market regulation. Conversely, in Deqing, where rural markets are still dominated by administrative principles, peasants, both as producers and consumers, find their choice of markets highly constrained.

Government regulations and trade barriers: Commodity war repeated

Direct government regulation is another important factor affecting China's rural market development. Skinner (1985a, 1985b) overlooked this factor in the reform period by assuming that the abolition of the People's Commune and the relaxation of government control restored economic forces. In fact, economic reform since 1979 has not altered the control mechanism created by the implementation of a planned economy. While combining institutional and entrepreneurial roles under the administrative-zone economy, local governments continue to intervene in economic activities. Such intervention has had a significant consequence on rural market development. This is particularly the case in Deqing, where economic activities are concentrated mainly in the official rural markets.

Trade barriers are a distinct form of government regulation seen in Deqing but not in Dongguan. They stem from the local government's attempt to block private trade in local specialities. Since the 1980s, the Deqing government has encouraged cultivation of pine resin, cassia bark and silkworm cocoons in an attempt to establish Deqing county as an export base for these products in Guangdong province. Unlike other sideline commodities, these three are categorised as special local products (*tu te chan*). The circulation and marketing of them are very different from other sideline goods such as pigs and vegetables. Special purchase stations have been established in every town to procure them from peasants. Private trading in these products is not prohibited by law. However, trading of these items in rural markets is very rare because local government strictly controls their flow. Moreover, it is equally difficult for external traders to purchase them directly from peasants.

Revenue raising is the major reason for direct government intervention in these local special commodities. During the 1990s, economic reform pushed local governments to implement a responsibility system for the acquisition of pine resin, cassia bark and silkworm cocoons. These changes were aimed at improving the efficiency of the old procurement system controlled by the central government through a scheme of sharing profits with subcontractors. However, local governments have not given a free rein in the trading of these items because they provide a major source of income for both local governments and peasants. Consequently, the procurement of a particular special local product is handled by one nominated subcontractor in each town. Interviews with peasants and government officials suggest that the sole subcontractor has a very close relationship with the local government, for example, in Decheng town, the subcontractor is a company established by the local government. This is a typical practice which demonstrates the intention of local authorities to retain the monopolistic trading of the pre-reform period. Moreover, despite the implementation of a responsibility scheme, as with the old procurement system, a town's procurement quota is assigned by the county government and is incorporated in the subcontractor's contract. Given the close relationship between the local government and the subcontractor, the two usually co-operate with each other to guarantee the assigned target is met. This safeguards both the subcontractor's profit and the local government's tax revenue.

Another major reason for government intervention is the variation in purchase price of special local products. Although this is set by the county government, some flexibility is allowed.[1] This allows each town's subcontractor to adjust the procurement price to increase profits and incentives. Accordingly, a price differential exists between towns. In 1988, for instance, the purchase price of pine resin was 90 *yuan* per 100 *jin* in Jiushi town, while that in Decheng town was 100 *yuan*. When compared with Deqing's neighbouring counties, the price differential was even greater. In 1998, the purchase price of resin in Yunan county and Gaoyao county was 200 *yuan* and 140 *yuan* per 100 *jin* respectively. Private traders offered an even higher purchase price of 250 *yuan* per 100 *jin*. The same price differential was observed in cassia bark. In 1998, the purchase price of cassia bark in Jiushi town was 330 *yuan* per 100 *jin*, while those in Decheng town and Nanjiangkou town it was 365 *yuan* and 400 *yuan* respectively.

Obviously, peasants wish to sell their products in places where higher purchase prices are offered. However, this jeopardises a town's procurement assignments and profits. Consequently, administrative measures (roadblocks, heavy penalties and threats) have been taken to stop the outflow of these products and so protect local government's interests. This is a unique feature of China's reform era that is not mentioned. This suggests that economic reform has not fully restored the prior position of market forces in rural market development which prevailed during the pre-communist period.

Administrative regulations have had a marked impact on trading and commodities flows and, therefore, on rural market growth. Discussions with peasants suggested that the most rigid situation was in Jiushi town. Peasants in Xiyan, Niutoutang and Yalizui villages disclosed that both the local government and subcontractor had blocked private exports of pine resin, cassia bark and silkworm cocoons. Roadblocks were set up on the town's only roadway to Decheng town and peasants' bicycles and tractors were checked. If peasants were caught, their commodities were confiscated, heavy fines were imposed, and their vehicles detained. In a small-scale peasant economy like Deqing, there was very little the peasants could do to prevent this intervention. Some had attempted to avoid the blockade by using the footpaths in the hilly areas. For example:

---

[1] Take pine resin for example. In 1995, its purchase price was 180 to 190 *yuan* per 100 *jin*. In 1996 and 1997, it was slashed to 140 and 90 *yuan* per 100 *jin* respectively. Purchase price of cassia bark also dropped from 600 *yuan* per 100 *jin* in 1994 to 330 *yuan* in 1998.

> Last year, Mr Li, who lives in Shanding village of Huilong town, 'smuggled' his pine resin to Decheng town, where the purchase price was higher than at the local purchase station (Figure 6.3a). Mr Li and his brothers left home at midnight to avoid roadblocks and checking stations. They travelled over 10 km on narrow rural roads and footpaths inside the hilly area to Decheng town and finally sold their pine resin at 150 *yuan* per 100 *jin*.

This case is very unusual because not many people were willing to take such high risks, instead, they have continued to complain and to resent the blockade. Since they have little choice other than to sell their products to the relevant local purchase station at a lower price, some peasants have stopped producing these specialised products in silent protest against the local government. This has further jeopardised rural market development.

Most peasants who rely on the pittance from pine resin, cassia bark and silkworm cocoon production, have been affected by restrictions on trading. This situation is best illustrated using a series of case studies. For example:

> Mr Lu lives in Shengtian village, a medium-size village with a population of 200. As he had six members in his family, the collective unit – Shengtian village – had allocated about 300 pine trees to him. All of his pine trees are mature enough to produce resin. According to Mr Lu, the collecting season of pine resin is between April and September of the lunar calendar. During this period, peasants can collect three or four times before they deliver their resin for sale. In 1997, Mr Lu's pine trees produced 7,000 *jin* of pine resin, all of which was sold to the government at the local purchase station located only 200 m from Shengtian village (Figure 6.3a). Although the purchase price in Decheng town was 30 *yuan* per 100 *jin* higher than at the local purchase station, Mr Lu did not sell his resin in Decheng because its outflow was blocked by local government. Such restrictions always caused tensions between peasants and the authorities and fighting usually erupted. However, peasants were worse-off in virtually all cases.

Peasants in Shatian village, which is located 2 km north of Shengtian, faced the same situation. Pine resin production is also a major sideline here, although as the village did not collectively own many pine trees only a small number of trees had been allocated to villagers. Take Mr Cui's case as an illustration:

> Apart from a small number of pine trees allocated by the collective unit, Mr Cui further subcontracted a considerable number of pine trees from his neighbour at one *yuan* per tree and had 200 trees in production. Being the

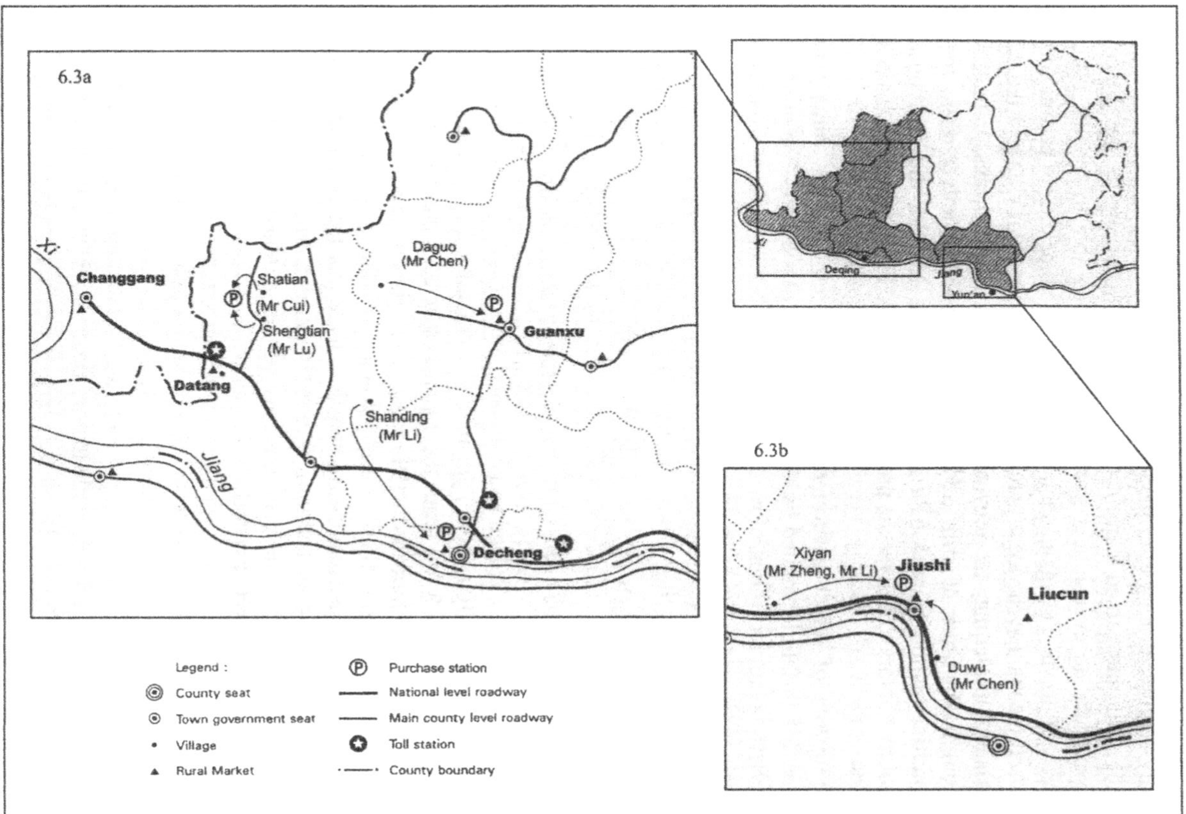

Figure 6.3 Peasant-market interactions under Deqing's single market structure and government regulation, 1997

> only labourer to look after the trees, in 1997 Mr Cui produced only 400 *jin* pine resin and earned a profit of 4,000 *yuan*. According to him, he could have made an even higher profit if he had been able to sell the resin in Decheng town. Yet again, such a movement was blocked, so he was forced to sell the resin at the local purchase station at Shengtian village. Like many peasants in Deqing, the bicycle was Mr Cui's major means of transport, though sometimes he still used the shoulder pole to deliver resins to the purchase station (Figure 6.3a). As a producer, Mr Cui did not consider renting tractors to deliver the products because he had to minimise his costs.

The impact of direct government intervention on Deqing's rural market development is further illustrated by the marketing of cassia bark. The production of cassia bark is similar to that of pine resin. Cassia bark and leaves are collected from cassia trees and provide the raw material for Chinese cinnamon, a popular spice in China. Many peasants have planted cassia trees in Deqing but the major bark production area is in Jiushi town. Unlike pine trees, almost all cassia trees are planted by peasants in their private plots. Although these trees take ten years to mature, bark and leaves are collected at the age of six and a higher yield is expected by the age of eight. For instance:

> Mr Zheng of Xiyan village had cultivated 5,000 cassia trees with an annual average yield of 300 to 400 *jin* of barks (Figure 6.3b). Although Xiyan village is situated 5 km from the Jiushi cassia bark purchase station, Mr Zheng sold his cassia bark there because it was the only place to sell bark in the town (Figure 6.3b). Like many peasants, Mr Zheng delivered his cassia bark to the purchase station by bicycle which was the cheapest mode of transport in Deqing. It took him 30 minutes. Another Mr Li of Xiyan village also sold his products in Jiushi purchase station, although he produced only 100 *jin* of cassia bark last year.

As with pine resin, peasants in Xiyan village knew that purchase prices differed between towns. Discussions with peasants suggested that the purchase price in Nanjiangkou, a neighbouring town located on the opposite side of Xijiang River, was 400 *yuan* per 100 *jin*, while that in Jiushi was 330 *yuan* per 100 *jin*. As local authorities had set up blocks on the quay, it was impossible to move cassia bark to other places. Moreover, peasants were also threatened not to sell their products to mobile traders. Consequently, despite the low price, both Mr Li and Mr Zheng thought the local purchase station was the only place to sell their cassia bark. In this situation, neither distance nor transport factors but rather government intervention determined the choice of markets. Such intervention prevents interactions between peasants and rural markets in other towns, and consequently, rural market growth is restricted as its hinterland is constrained by artificially imposed barriers.

Apart from pine resin and cassia bark, local governments have also interfered with the marketing of silkworm cocoons, which must also be sold at the designated purchase stations. Unlike pine resin and cassia bark, not many Deqing peasants still keep silkworm cocoons because mulberry trees, on whose leaves silkworms' feed, grow only in flatland areas. In Deqing county, the major silkworm cocoon production area was restricted to Jiushi town. In 1997, there were 1,540 *mu* of mulberry trees in the town, accounting for over 19 per cent of the county's total, and silkworm cocoon production was 83 tonnes, ranking Jiushi second among the 14 towns in the county (DQSB, 1998, p.45).

Dúwu village of Jiushi town was a place in which all households were once involved in raising silkworms. This changed when a considerable area of farmland and private plots belonging to peasants were requisitioned for roadway construction. Mr Chen was one of the very few who insisted on keeping silkworms. Land requisition had less of an impact on his household because he still owned a 2 *mu* plot for growing mulberry trees, apart from these, extra space was also needed for a special room for keeping silkworms. However, unlike pine and cassia trees which took a long time to mature and produce resins and bark, the silkworm cocoon's growing period was only 25 to 27 days. The short growing period allowed Mr Chen to harvest seven yields of silkworm cocoons each year. April to October is the main growing season, the other months are too cold. Silkworms were raised in round bamboo trays one metre in diameter, layers of trays were put in special rooms where sunlight was screened and light bulbs installed to keep them warm at night.

Silkworm larvae were available at the silkworm purchase station which was located in the town-government seat of Jiushi. Mr Chen usually bought 1 to 1.5 *liang* larvae for each yield.[2] According to Mr Chen, silkworm raising was a labour-intensive job because they had to be fed four times a day, i.e. morning, afternoon, evening and midnight, and mulberry leaves were collected before each meal to ensure their freshness. Moreover, silkworm excrement needed to be removed frequently and lime powder was required for fumigation. After 15 days when silkworms had grown to a considerable size, they were transferred to straw cocoon nets to spin cocoons, these were mature and ready for sale in another 10 to 12 days. Usually, one *liang* of silkworm larvae yielded 50 to 60 *jin* of cocoon.

The purchase price of cocoons was still controlled by Deqing's county government, despite the deregulation of other items during the reform era. In

---

[2] Ten *liang* is equal to 1 *jin*, which is about 0.5 kg. They are all units of weights that peasants use.

Deqing, the purchase price had decreased since the 1990s, making silkworm raising no longer a high income sideline.

> According to Mr Chen, several years ago the purchase price of silkworm cocoon was 8 *yuan* per *jin*. By June 1998, it had dropped to 5 *yuan* per *jin*, and in July 1998 had further decreased to 4 *yuan*. Consequently, Mr Chen's income had been slashed from 500 *yuan* per yield to 240 *yuan* per yield, yet production costs had increased. As stressed by Mr Chen, only marginal profits were expected when the purchase price was around 6 to 7 *yuan* per *jin*. Increases in production costs and a decline in purchase prices had wiped out the marginal profit of peasants along with the incentive to keep silkworms. Already, many in Dúwu village had given up silkworm cocoon production; although Mr Chen had persisted, he kept far fewer. He left the job of tending them to his father while he worked as a construction worker in town.
>
> Although Dúwu village was located 1 km from the Jiushi cocoon purchase station, distance was not the major factor that determined Mr Chen's choice of market. He was obliged to sell his cocoons at the local purchase station because there was no other choice in town (Figure 6.3b). However, the purchase price was higher in Liudou, a neighbouring town situated on the other side of Xijiang River. Yet he had never thought of selling his cocoons there because local authorities had set up blockades. Also rogues were employed by the subcontractor to threaten peasants not to export their products. If peasants were captured at the roadblocks, all cocoons would be confiscated and a fine imposed. For the same reason, it was rare to have mobile traders from other places purchasing cocoons in local villages.

Thus, the flow of cocoons had been strictly controlled by the local government.

Trade barriers were not imposed on other sideline products but it was clear that administrative intervention was a major obstacle preventing the opening-up of Deqing's economy. This invisible wall – government intervention – also undermined the benefits brought about by transport improvements in recent years. Unlike Deqing, however, direct government intervention in production and marketing of agricultural products was weak in Dongguan. It appeared that Dongguan's open economy and strong industrial development had wiped out local government interest in controlling agricultural production and commodity flows. The extent of local government control in Dongguan was beyond the scope of this study, nevertheless, situations similar to those in Deqing were not found in Dongguan.

Apart from artificial trading blockades, direct administrative intervention was also observed in the production of speciality items, resulting in their decline. The collection of pine resin is a good illustration, because it was big business in Deqing. Not only did local governments attempt to monopolise procurement, but administrative villages – village-level administrative units – also interfered in pine resin production. The most common strategy was to retain all the collectively owned pine trees and not allocate them to peasants, so while many peasants produced pine resin, they did not own any trees.

The case of Daguo village within Guanxu town is a good example. Peasants there had to rent pine trees from the collective unit because they were owned by the natural village. Mr Chen was one of the peasants living in this small village.

> According to him, rents varied between different villages. In Daguo, rent accounted for about 25 per cent of the income of peasants from pine resin. While Mr Chen earned 2,000 *yuan* from pine resin in 1997, 500 *yuan* was paid to the collective unit for rent. Moreover, a licensing system was about to be implemented, under which those involved in pine resin production would have to register with the related authority. Mr Chen was very disappointed with this system because more levies were likely to be imposed on peasants. He thought it was another way for local governments to milk profits from pine resin production.

Consequently, the earnings of peasants from pine resin are expected to diminish.

Another form of roadblock: Toll stations

Toll stations represented another form of direct government intervention which affected market development during the reform period. They are found in Deqing and Dongguan and have had a varying impact on rural market development in the different local contexts. Unlike intervention in the trading of local special products, local governments did not intend to use toll stations to prevent the movement of people. However, these stations have, unexpectedly, restricted the choice of markets by preventing the free journeying of people between towns. As interactions between peasants and rural markets were restricted, market development was influenced adversely. Toll stations were established by the local government and managed by the Roadway Bureau. If transport factors are seen as the 'invisible hand', toll stations are the 'invisible walls' that constrain the flow of people, vehicles and even commodities. The toll station has become a common form of control in Deqing and Dongguan – in Deqing county there were four in 1997 and although it was difficult to determine their precise numbers in Dongguan, two stations were located in Wangniudun town.

Toll stations were a direct result of dual administrative and entrepreneurial roles performed by local government. As the Deqing government took out loans to pay for road construction, the toll payment represented the only, as well as the fastest, way of repaying debt. This method has been prevalent during the reform period which has reinforced the economic decision-making of local governments.

The Roadway Law imposed in 1998 has legalised this practice.[3] Toll stations have been allowed on county-level roadways – jointly developed by local and foreign capital – and have become a prime source of local government income which explains why so many have been established (Chung, 2002). Some illegal ones have also been created. As toll stations were generally set up at town or village borders, the movement of people between towns was reduced, and market development influenced adversely.

Thus, the positive relations between improved transport routes and market development highlighted by Skinner have not occurred in Deqing. Moreover, the negative impact of toll stations was reinforced by the county's single market structure and its simple road system. Since almost all towns in the county had only one rural market, if peasants bypassed it they had to cross the town border to visit markets elsewhere. By erecting toll stations the flow of people and commodities was easily inhibited because all paved roads in Deqing were part of the county's main transport artery. In contrast, although there were more toll stations in Dongguan, their impact on market development was less because of the existence of both village-level and town-level markets which gave peasants more choices within their hometowns. Moreover, while toll stations were established in Dongguan, toll payments were not a major concern because peasants were wealthier there than in Deqing. Table 6.1 shows the toll rate of Wang-Hong roadway toll station in Wangniudun.

In Deqing, the impact of toll stations on rural market development is aptly exemplified by the decline of Datang market in Huilong town. This was the sole market in Huilong situated 2 km from the county border – a traditional site which had attracted peasants from surrounding towns and counties. Prior to the opening of the toll station in the 1990s, the market drew peasants from Changgang towns in Fengkai county. Since its establishment at the border, an 'invisible wall' has been created and interactions between Datang market and places outside Deqing have been completely cut off. Tolls required for various types of motor vehicle are shown in Table 6.2. This 'invisible wall' has had a devastating effect on activities at Datang market which was formerly a thriving economic centre.

As examples, Shegui and Jiangbianwei were two villages located in Changgang town that were linked to Datang and Changgang rural markets by a roadway. The establishment of a toll station at the border has discouraged peasants in these two villages from visiting Datang market. For instance:

---

[3] The Roadway Law was passed in People's Congress in July, 1997 and took effect on 1 January, 1998. The law authorises local government to establish toll stations and collect tolls within their administrative jurisdictions. Generally, toll stations are allowed every 40 km and 20 km in plains and hilly areas respectively.

Mr Yang's family was one of 23 households living in Jiangbianwei village. Although the village was located 2.5 km to the west of Datang market, Mr Yang always visited Changgang rural market situated 7 km to the west of the village (Figure 6.4). He did not visit Datang market simply because of the toll station. Discussions with peasants in Changgang suggested that although local bicycles were exempt from paying tolls in principle, policies were inconsistent and peasants were sometimes caught by the authority and penalties imposed. To avoid possible trouble, Mr Yang chose to visit the local market rather than cross the border. Using a bicycle as his major transport mode, Mr Yang bought non-staple foods, clothes and other daily necessities in Changgang rural market.

Table 6.1 Toll rate in Wangniudun, Dongguan, 1997

Type of vehicle	Toll rate (in *yuan*)
Motor-cycle and sedan	2
Passenger vehicles at/below 20 seats and trucks at/below 2 tonnes	5
Passenger vehicles between 21-50 seats and trucks between 2-5 tonnes	10
Passenger vehicles at/above 51 seats and trucks between 5-50 tonnes	15
Passenger vehicles and trucks over 51 tonnes, and container trucks	25

Table 6.2 Toll rate in Deqing county, 1997

Type of vehicle	Toll (in *yuan*)
Motor-cycle	3
Sedan, small vehicles	10
Van, medium-size vehicles	20
Truck, large vehicles	30
Container truck	35

Mr Bian, who resided in Shegui village, 4 km and 5.5 km from Changgang and Datang rural markets respectively, chose Changgang market instead of that at Datang (Figure 6.4). Although he said distance was the prime reason for his choice, he also mentioned that the toll station at the border was an equally important factor. It has discouraged people, including himself, from visiting Datang market.

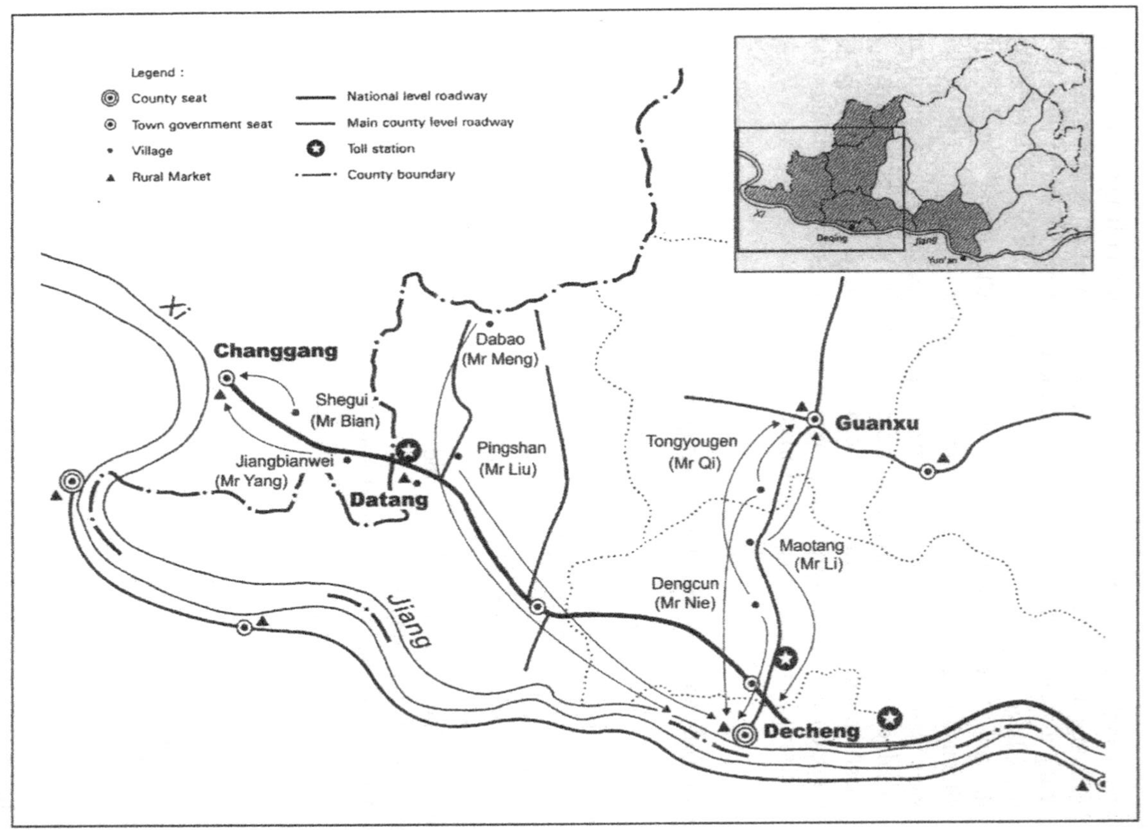

Figure 6.4 Toll stations and their impact on Deqing's peasant-market interactions, 1997

Similarly, peasants from Huilong town rarely crossed the border to visit the Changgang market. Because of the difficulty of access from neighbouring villages, Datang market also remained quiet – its turnover value in 1997 was 2.6 million *yuan*, the lowest in Deqing county; its annual number of visitors, only 0.29 million, was also the smallest.

The Datang example contradicts Skinner's general thesis that improved road routes will increase the value of rural market turnover. Since a paved road linked Datang with other towns, the stagnation of its market could not be seen as a consequence of poor transport routes. Also, it contradicts Skinner's (1965a, p.220) argument that the demise of standard markets can follow improvements in road links to higher level markets. In the Datang case, the need to comply with government regulations at the toll station outweighed the benefits of a good road. Worst of all, not only did peasants in Changgang town rarely cross the border, but Huilong peasants also ignored Datang market. For example:

> Mr Meng lives in Dabao village, the remotest village in the study area. Located in the north hilly border of Huilong town, the village is situated 9 km and 25 km from Datang and Decheng rural markets respectively. Despite the long distance, Mr Meng and other villagers bought clothes and shoes in Decheng rural market (Figure 6.4). According to Mr Meng, his choice was due to the larger size of market and greater variety of commodities, compared with Datang. The long distance between Dabao village and the market prompted Mr Meng to rely on local bus services – referred to as 'small wheels'. Since 'small wheels' were privately owned, the fare was 7 *yuan* and travel time was seventy-five minutes. In fact, Mr Meng was not a big farmer with a stable income. Cultivating less than 1 *mu* of paddy field and earning a pittance from pine resin production, Mr Meng is one of many small peasants in Huilong town. Owing to the decline of the local market, Decheng rural market had thus become the preferred market-place for Mr Meng. However, low income, long distance and high bus fares have constrained Mr Meng's shopping pattern. Last year, he visited Decheng rural market only once, buying a new shirt for the Chinese New Year.

Mr Liu of Pingshan village also provides important insights:

> Living 18 km to the west of Decheng rural market, Mr Liu also purchased clothes and children's wear in the market. Mr Liu said he was not a choosy consumer but the local rural market – Datang – did not provide the commodities he needed. Decheng market has become therefore the nearest suitable market, despite a travelling time of over 30 minutes and bus fare of 5 *yuan* (Figure 6.4). Like many peasants in Huilong town, Mr Liu said

> family income has determined his shopping pattern. As a result, he visited the market only twice last year.

As local peasants bypassed Datang, the market remained quiet on market days, with trading hours now shortened to one to two hours. Since Datang market was established under administrative principles, it will not die 'naturally' despite the decreased number of visitors.

The decline of Datang market has been recognised by Deqing county's ICMB. However, the authority considered that the toll station was not an issue in its decline, believing instead that the sparse population of Huilong town was the prime factor. Despite the downturn of Datang market, a new one was planned in Huilong's town-government seat. If a sparse population has caused stagnant market conditions in Huilong, the establishment of another market in the town will prove to be a poor decision. Indeed, local peasants did not have high expectations about the new venue. Despite construction work begun in 1998, the establishment of this market was eventually abandoned in 2000. Thus, there is no doubt about the negative impact of the toll station on the choice of market by peasants and on Deqing's overall rural market development.

In Dongguan, toll stations appeared to have had little impact on rural market development. A major reason is that Dongguan's rural market system has always been more advanced and diverse than that of Deqing. In Wangniudun town, there were eight rural markets and five of them were at village-level. They provided an ample choice of foodstuffs, clothes, household articles and electrical appliances, so peasants seldom crossed the border to other venues because a full range of commodities was available within their own town. For example:

> Mr Dui of Duwuxin village purchased foodstuffs, footwear, children's wear and other daily necessities at the village-level and town-level markets (Figure 6.2). As Mr Dui had a motorcycle, he travelled 3 km to visit the town-level market two to three times a week. According to him, he seldom visited venues outside Wangniudun town because he was satisfied with the variety and quality of commodities locally. Last year, however, Mr Dui made one visit to Guancheng, the city government seat, to buy an imported colour television set.

Clearly, the advanced market system in Dongguan had neutralised any negative impact caused by the toll stations.

That toll stations were responsible for the decline in Deqing is further demonstrated by the fact that their abolition elsewhere has produced the reverse effect. This is illustrated by the example of Guanxu rural market located 13 km to the north of Decheng central market, the largest in Deqing. Although linked by a paved road, peasants in Guanxu seldom visited the Decheng market because a toll station had been established on the only roadway connecting the two towns. Since its abolition peasants' visits have become more frequent. For example:

> Mr Qi, who lived in Tongyougen village of Guanxu town, had a motorcycle but seldom visited Decheng market because of the toll station. He said the toll station had severed local people's connections to the Decheng market because many of them could not afford the toll payment. After the toll station was abolished in early 1998, Mr Qi travelled frequently to the Decheng market (Figure 6.4). He said the elimination of the toll station was of significant consequence to him and other motorcycle riders.

Since peasants can now travel freely to Decheng market, their trips have become more frequent.

The abolition of the toll station has also attracted many Guanxu peasants, like Mr Li, to the Decheng central market:

> Mr Li lives in Maotang village, a medium sized village with a population of over 500. As one of his sons has been a migrant worker, family income had been quite stable. Several years ago, Mr Li had spent 30,000 *yuan* rebuilding his house. Although he was relatively wealthy, he never visited Decheng market before the abolition of the toll station. Guanxu market, which was located 6 km from Maotang, was his sole choice for buying clothes and daily necessities (Figure 6.4). While cycling to Guanxu market took him about 40 minutes, Mr Li said the distance was acceptable. After the removal of the toll station, Mr Li switched to the Decheng market, which was larger in scale and had a greater variety of commodities than Guanxu. Although Decheng market was located 7 km from Maotang village, Mr Li still insisted on cycling, though sometimes he chose to take the recently introduced public bus services. Mr Li was pleased to see these changes and he thought the 2 *yuan* bus fare was acceptable. As a result, he visited Decheng market more frequently than before to purchase clothes, footwear, daily necessities and non-staple foods.

The situation of Mr Nie, of Dengcun village, also illustrates the positive effects of abolishing the toll station:

> He did not visit Decheng market when the toll station was in operation. Although Dengcun village was situated 9 km to the north of Guanxu rural market, it was his only choice for buying food and daily necessities. The abolition of the toll station has enabled him to visit a bigger and closer market, which was located 4 km to the south of Dengcun village. Although the bus fare was only 2 *yuan*, Mr Nei preferred cycling because he thought 4 km was only a very short distance. While cycling to Decheng market took Mr Nei 20 minutes, he did not visit the market regularly. Like many peasants in the county, he thought income had significantly affected his

> choice of market and shopping frequency. Nevertheless, it was the abolition of the toll station that let him choose between a wider range of markets (Figure 6.4).

Consequently, Decheng market has become busier because peasants are no longer restricted by government regulations from visiting it. Moreover, the abolition of the toll station has also increased the efficiency of the road (see Chapter 5). The positive impact on rural market development brought about by road improvement has been revived.

Case studies of Deqing and Dongguan have highlighted that government strategies included direct intervention through price controls and roadblocks, and indirect intervention through the establishment of toll stations. However, the impact of all these factors has varied depending on local contexts. While roadblocks and price controls have greatly restricted producers in Deqing from marketing local products, the consequences have been less significant in Dongguan. Toll stations have created unexpected barriers to the choice of markets in both places. However, given the variations in economic development and market establishment, the effect of government regulation on rural market development has been more substantial in Deqing than Dongguan.

It is worth mentioning the significance of administrative boundaries at the end of this discussion. The use of administrative principles to establish markets, the attempts by local government to block private trade in special products and the establishment of toll stations have implicitly referred to the functions of borderlines at town and county levels. Indeed, administrative boundaries represent another icon of government regulation in China. While Thuen (1999, p.740) regards national borders as "dividing lines between different regimes for the ordering of right over space", local governments' administrative boundaries in China can be understood as dividing lines between local market systems. During the pre-reform period, when rigid administrative measures were used to regulate economic activities, economic hinterlands of local market systems were bound basically by administrative boundaries of town and county (i.e. a political-cum-economic system). In addition to the household registration system and a centralised resources allocation system, commodity and people flows were highly restricted. Since all interactions were directed by vertical channels, i.e. orders and plans, aligning with the administrative system (Chung, 1994), marketing activities were confined to a particular administrative territory with very little cross-border interactions.

Since 1978, the malfunction of the household registration system and the dismantling of the centralised resource allocation system have generally undermined the importance of administrative boundaries. Thus, the direct impact of borderlines on restricting people and commodity flows has weakened. Nevertheless, the establishment of roadblocks and toll stations on county borders have reactivated the function of administrative boundaries as dividing lines between market systems over space. This is particularly the case in Deqing – a semi-closed economy in a hilly area. Moreover, the rejuvenation of borderlines' administrative significance has reinforced the political-cum-economic structure of Deqing. Fieldwork in this area suggests the same development in Deqing's neighbouring counties, such as Yunan and Fengkai, where a similar physical landscape, socio-economic and political conditions and accessibility are shared. As a result, on a regional scale, a cellular spatial structure based on county is produced. Within each 'cell' is a small network with highly comparable, if not identical, economic structures. Ironically, Deqing government's effort to improve accessibility of the county has alienated itself from neighbouring economies.

A cellular structure is also observed in Dongguan, although it has little relation to rural marketing activities. This is because the Pearl River Delta in general and Dongguan in particular have been transformed from agricultural to industrial and manufacturing bases. Thus, rural marketing activities are no longer a priority concern of local governments and hence not a major target for regulation. However, the political-cum-economic system has persisted, despite the 'invisible hand' being more influential in the Delta region. Administrative principles have built the skeleton of Dongguan's market system and government regulations persist. However, unlike Deqing, state governance, as argued by Tang and Chung (2000), has shifted from direct surveillance to indirect measures.

The significant impact of administrative factors in rural market developments in both Deqing and Dongguan is not uncommon. A similar situation has been observed in the Sudan in Africa where marketing activities are constrained by state monopoly, and wage and price control (Lado, 1988). However, government regulation in China has been more dominant and decisive. It could be regarded as an 'invisible wall' that blocked movements of people and commodities between towns and counties and which has wiped out the benefits created by transport and economic improvements. This has been the case mainly in Deqing county, as market forces were not as well established as in Dongguan.

In short, the significance of administrative factors has not diminished during the reform era. Skinner has addressed their importance but he also overplayed China's market liberalisation during this period. The cases of Deqing and Dongguan have provided strong counter-arguments to Skinner's rather parochial views on the importance of government regulations. In addition, variations between Deqing and Dongguan have suggested that China's rural market development is much more complicated than envisaged in Skinner's analysis.

Chapter 7

Deqing, Dongguan and Spatial Variations

Previous discussions of Deqing and Dongguan demonstrate variations rather than similarities. Despite Deqing and Wangniudun of Dongguan having their own intermediate market system, both places have distinctive types of local markets. The former exhibits a simple structure with a high degree of government intervention while the latter is more open and sophisticated with a relatively low degree of regulation. In Deqing, marketing and shopping patterns demonstrate that sales and consumption are highly concentrated in rural markets as a result of the county's relatively poor retail network and road system. In contrast, Wangniudun has a diversified shopping and marketing pattern, made possible by its well-developed transport network and the relatively high income of its peasants.

Spatial variations are addressed by Skinner through his topographic models and the regional framework. The former, as noted in Chapter 2, attempts to address the differences in marketing pattern and structure in mountainous and plain areas. They present a strong sense of environmental determinism because physical environment is regarded as a determinant for social and economic activities. This perspective has limited the models' explanatory power. Thus, they are insufficient to explain the differences between Deqing and the Wangniudun market of Dongguan and their relationship with neighbouring market systems.

Unlike mountainous and plain models, the regional framework addresses spatial variations by reference to the concentration of trade and socio-economic development. The foundations of this framework are rural market systems. They are regarded by Skinner as forming the basic level of a regional hierarchy, in which these systems are connected horizontally to similar local systems, and vertically to higher level systems. A total of eight macroregions have been formed in China. Each region features a "system of cities that provide[s] its skeletal structure" and is further divided into cores and peripheries (Skinner, 1977b, p.212; 1994). Core and periphery are initially defined by using:

> population density as the major indicator of resource concentration on the grounds that it is a major component of demand density and correlates strongly with agricultural productivity both as cause (labour inputs) and as effect (the carrying capacity of the land). (Skinner, 1977c, p.282)

According to Skinner (1994, p.49), each macroregion's core and periphery have a similar level of development, though some of them are situated in China's inland and border areas.

The empirical validity of Skinner's concept is examined by Sands and Myers (1986). In considering six rural variables – population density, cropping density, total grain surplus, total sown area, proportion of irrigated farmland and proportion of farming households – they question the value of his notion because their findings show no marked differences in rural activities between macroregions. Also, they argue that Skinner's core-periphery hypothesis has oversimplified the situation during the early twentieth century. Using the same indicators to examine concentration of resources in macroregions, Sands and Myers (1986, p.737) argue that Skinner's idea does not provide an effective tool for understanding Imperial China because these concentrations varied between the cores and peripheries of different macroregions.

Skinner's (1994) recent work argues that the legitimacy of the core-periphery notion has persisted during the reform era. However, here he adopts a gradual developmental model to explain spatial variations between core and periphery to accommodate the complexity of economic and social development in China during economic reform. Using the Lingnan macroregion as a case study, Skinner (1994) reclassifies it into seven zones in accordance with economic, educational, welfare, and demographic variations.[1] From the centre to the rim, these zones include an inner core (Zone 1), outer core (Zones 2 and 3), near periphery (Zones 4 and 5) and far periphery (Zones 6 and 7). In terms of development, according to Skinner (1994, p.28), "Lingnan's Zone 1, the inner core, is comparable to the Four Dragons, whereas Zone 7, the far periphery, resembles Nepal".

In this scheme, Deqing would be located in the outer core (Zone 2) and Dongguan in the inner core (Zone 1). There is little doubt about Dongguan's central location and prosperity. Conversely, given Deqing's stagnant economic and social development, it is surprising to find that the county is included as a core area in Skinner's scheme. Since the rural market system in Deqing is rudimentary and its economy semi-closed, one can only speculate about its counterpart in the far periphery areas (Zone 7). Without comparing a marketing district from the far

[1] Skinner (1994, pp.33-47) analysed 20 variables. They are the level of mechanisation, irrigation, tractor ploughing, electricity use, grain yields, meat production, gross value of agricultural output, agricultural labour force, industrial labour force, size of largest town, gross value of industrial and agricultural output, illiteracy rate, junior-middle schooling level, infant mortality, farm income, child-woman ratios, general fertility rates, age structure, household size, and migration.

periphery with those of Deqing and Dongguan, it is hard to validate Skinner's classification. His claim has to be the topic of further research.

Skinner's (1994) recent modification of his original work is still problematical, despite various inclusions to strengthen his argument about China's core-periphery structure. The fundamental problems stem from his failure to include political and policy factors, alongside the economic, demographic, and transport issues which he has considered. The differences between Deqing and Dongguan result from the varied effects of market forces and government regulations, demonstrating that a shift in Chinese development philosophy during the reform era has produced considerable unevenness in market development. The socialist ideal of equality between regions and between urban and rural areas during Mao's period has been replaced by Deng's 'let a small number of people get rich first', which places greater emphasis on market forces to achieve rapid economic growth. Under this ideology, a widening developmental gap between coastal and inland and core and periphery is tolerated (Fan, 1997). Scholars, such as Hu (1998), argue that unequal growth is inevitable when China is in transition from a planned economy to a market economy.[2] Like some Chinese leaders, Hu seems to believe that trickle-down effects will finally bring prosperity to the whole of China.

There is no doubt that while market forces have boosted economic development they have also increased regional inequality during the reform era. Paralleling the shift in China's development philosophy has been the adoption of a coastal-oriented strategy whereby foreign investment has been pumped into coastal areas such as Dongguan. The establishment of special economic zones, open coastal cities, economic and technological development zones and open coastal economic zones has given preferred status to specific places. Benefits of this include exemption from customs duties, industrial and commercial taxes, local income taxes and the right to retain local foreign exchange earnings. In contrast, inland areas, like Deqing, have not been covered by these policies, serving to widen the gap between these areas. The variation in foreign investment input in Deqing and Dongguan illustrates the disparities created by preferential policies (see Chapter 4). Foreign investment, however, is missing in Skinner's (1994) data sheet, though he claims his figures have incorporated the effects of these policy changes.

The shortcomings of the core-periphery concept in explaining variations in post-1979 China are further demonstrated by examining changes to the administrative hierarchy. Focusing on Imperial China, Skinner suggests a

---

[2] In contrast to Hu, Wang and Hu (1999) argue that the unequal nature of China's regional development is avoidable. This is because the inequality is generated by preferential policies to specific places. Abolition of these policies may generate a relatively balanced growth between different regions.

correlation between administrative and economic systems in core and peripheral areas and argues:

> Cities whose economic centrality was exceptionally high tended to be capitals of extra-large prefectures with extraordinarily broad spans of control......Span declined steadily and regularly with the capital's economic centrality until......cities of exceptionally low economic centrality are seen to have been capitals of very small prefectural-level units with extraordinarily narrow spans of control (Skinner, 1977c, p.312).

In other words, high-level administrative units, such as prefectures, were situated in core areas and low-level administrative units – *zhou (chou)* and *ting (t'ing)* – were found in peripheral areas. Since 1949, *zhou* and *ting* have been abolished and replaced by town and township as the lowest levels of the administrative hierarchy.

The case studies of Deqing and Dongguan show that there is little correlation between the centrality of location and the span of control. Dongguan enjoys a central location but its span of control is similar to that of Deqing. Both are encompassed by administrative boundaries. Town and township are both found in Deqing and Dongguan. Moreover, since towns in both places are at the same administrative level they have the same span of control, irrespective of whether they are situated in the centre or the edge of Guangdong Province. Thus, a place's span of control is defined by its administrative status and not by location. This is particularly significant in China where the administrative hierarchy takes precedence over the economic hierarchy.

The weak correlation between location and span of control is generated by the institutional reform of city-leading-counties (*shi guan xian*) since 1982 which has re-arranged China's spatial organisation. The power of prefecture-level cities has been expanded by putting counties and towns, regarded as rural, under their jurisdiction. In so doing, urban and rural areas have been placed under one administrative unit. This extension of power, however, does not imply that the core area has also expanded. In fact, it indicates that a rapid urbanisation process has taken place through government regulation.[3] Many new prefecture-level cities, regarded as urban, are now set up in non-central locations. Guangdong was one of the first three provinces to implement the city-leading-counties system. Under the new arrangement, Zhaoqing city, the prefecture-level city to which Deqing is administratively subordinated, has been established in the western border of the

---

[3] Another measure to accelerate urbanisation is the change of establishment criteria. Yang (1996) gives a detailed description of these. See Lin and Ma (1994) for a description of the urbanisation process in Guangdong.

province. Moreover, designated towns have been set up in both central and periphery locations. Many of them in Guangdong are local economic and transport hubs. This is particularly the case for those acting as county government seats (Lin, 1993).

The limitation of the core-periphery concept is further illustrated by variations in government control which are reflected in Deqing and Dongguan. If, as Skinner suggests, the span of control is broad in core areas and relatively narrow in peripheral ones, the degree of government control should also be tight in the core area and loose at the periphery. In fact, Deqing and Dongguan demonstrate the opposite case. The impact of toll stations, existence of trade barriers and direct regulation of agricultural production in Deqing highlights the fact that the border county suffers from a greater degree of government control than Dongguan.

In studying China's urbanisation process, geographers attempt to explain why spatial variations exist in the degree of government control. Employing an economic imperative, Chan (1992) suggests that tight control in urban areas stemmed from the high cost of urbanisation. During the pre-reform period, a series of measures was implemented to minimise the cost of urban development. In all cities, these included tight management of migration, a rationing system for all necessary consumables, full employment, suppression of both incomes and consumption of urban workers, and limitation on the growth of the services sector. As argued by Chan (1992), the aim was to 'economise' the cost of urbanisation without hampering the pace of China's industrialisation.

Tang (1997) further exposes how the tight control of urban areas has been brought about by China's 'shortage economy', characterised by expansion drives and investment hunger.[4] To prevent the expansion of cities, created by industrial activities, the state had to limit their growth – a situation absent in rural areas which are dominated by agricultural production.

The persistence of 'shortage' conditions during the reform era has been dependent upon the level of market liberalisation. When central planning resulted in resource shortages, state regulation was reinforced. The introduction of market forces during the reform era has proved to be an effective tool in breaking down this vicious circle. Another measure used to combat shortages is to increase local government's autonomy by decentralising economic decision-marking power. Both courses, however, have been ineffective in improving Deqing's situation. Obviously, Deqing, which is located on the provincial border and enjoys no preferential policies, is under less regulation from market forces than Dongguan. The relatively poorer natural and economic conditions present in Deqing have

---

[4] The concept of a 'shortage economy' was originally presented by Kornai (1980a, 1980b). It provides an excellent tool to analyse the economic situation in planned economies like China.

resulted in the continuing of resource shortages, despite the decentralisation of power to local government. For Deqing, faced with heavy fiscal pressure stemming from the termination of direct subsidies from the state, the increasing cost of bureaucracy, and working targets set up by governments at higher levels, government regulation is probably the best way to allocate local financial resources to selected sectors and protect them from economic risk. In contrast, Dongguan has enjoyed a massive influx of domestic and foreign capital as a result of various preferential policies. Decentralisation of economic decision-making power has enabled Dongguan to make use of a golden opportunity to develop into one of the coastal region's most prosperous places. With the loosening of central control, government surveillance has changed from a direct to an indirect form of intervention.

Variations in government control have also been examined by political scientists from a different perspective. Focusing on administrative structures and personal appointment procedures, Edin (1998) suggests that tight control at the local level is a double function of fiscal incentives available to local government and personal interest of local cadres. As noted, these fiscal incentives are driven by financial pressure generated by economic reform. The prospects for promoting local cadres are determined by their achievements, evaluated ostensibly on the basis of both their economic and non-economic performances – usually the former takes precedence (Ho, 1994). To achieve their targets, local cadres use their authority to concentrate effort and resources on monitoring selected sectors and industrial enterprises (Edin, 1998). Consequently, more local resources and enterprises are put directly under local government control.

Government control of local resources does not automatically generate economic growth. Where economic objectives are hard to achieve in hilly areas like Deqing, local cadres pay more attention to non-economic goals including the implementation of unpopular policies such as birth control and revenue collection (O'Brien and Li, 1999). This explains why social control persists in areas of poor economic performance. In contrast, social control in economically prosperous areas, such as Dongguan, is relatively loose because local cadres expend most of their efforts on economic issues.

The situation in Deqing and Dongguan suggests the core-periphery concept is inappropriate for explaining spatial variations during the reform period. The notion is fundamentally economic but not political/administrative in nature. It explains spatial patterns in a general manner but fails to address disparities caused by structural forces. Although these matters have been criticised by Wang and Hu (1999, p.207) as not being in conformity with the socialist ideals, they are crucial for understanding China's spatial development during the reform period. During the debates over a new regional geography, Passi (1991) suggests the concept of "institutionalisation of regions" to understanding the transformation of space. According to him, institutionalisation occurs through boundary making, the formation of governing agencies that embedded in that territory, and the reproduction of "regional identity" (Passi, 1991, p.244). This perspective has not only proposed a new way to understand the formation of regions, but also the variations between them. It is to these conditions that the discussion now turns to the urban-rural perspective.

Urban-rural perspective: New meaning in a new context

The shortcomings of the core-periphery notion in explaining variations between Deqing and Dongguan invites reconsideration of the urban-rural concept, which has been the subject of considerable scholarly debate in recent years. McGee (1998, p.471) has suggested that the urban-rural dichotomy is losing "its utility as a heuristic concept and a policy paradigm". He further argues that "the concept of rural-urban division based on spatial demarcation is merely an artificial one" (McGee, 1998, p.471). The reason for this is because the dichotomy has created difficulties in placing definitive boundaries around the concepts of 'urban' and 'rural', particularly in cases where the development of globalisation and the global imperative of production, investment, and trade have increasingly integrated regional economic activities. Koppel (1991, pp.60-61) believes these activities have not only led to the influx of private capital into urban centres, but have also redefined the rural service markets involving land and labour. Consequently, urban and rural areas are no longer seen as two separate, antagonistic areas.

The other rationale behind McGee's argument is his study of functionally integrated zones and megapolitan regions in Asia's developing countries. Based on his empirical studies of the urban transition, McGee argues that functionally integrated zones, which combine both agricultural and industrial activities, are emerging. He uses the term *desakota*, the Indonesian word meaning village (*desa*) and town (*kota*), to denote these integrated zones (McGee, 1989, pp.93-94). According to McGee (1989, p.96), *desakota* is a region which has "no clear cut division between rural and urban relations". Functionally, it is "an intensive mixture of agricultural and non-agricultural activities that often stretch along corridors between large city cores" (McGee, 1991, p.7).

By applying the *desakota* concept to Asian countries, McGee and his followers argue that functionally integrated zones have appeared in Taiwan, Japan, Indonesia and various parts of China (McGee, 1989; Zhou, 1991; Johnston, 1994; Lin, 1997b; Wang, 1997). The emergence of *desakota* implies cities and their rural counterparts are no longer separated and disparities between them are receding. In turn, urban-rural interactions have been reinforced and both areas have become increasingly integrated in a "web of transactions" (McGee, 1998, p.485). This development suggests urban and rural areas have been integrated by a regional network. Within this network, the urban and its rural counterpart are seen as being well connected by transport systems allowing frequent interactions of people, commodities and capital flows. These linkages are said to narrow the development gap between the city and countryside.

McGee's argument is not without its supporters. Koppel (1991) and Koppel and Hawkins (1994) also question the legitimacy of the urban-rural dichotomy in understanding current developments in Asian countries. Koppel and Hawkins (1994, pp.4-7) suggest that a crucial step in understanding the complexities of rural labour processes in Asian countries is to reject the primacy of the urban-rural dichotomy. Koppel (1991, p.48) regards the dichotomy as a "deceptive dilemma" of two misleading alternatives. He criticises studies employing this perspective as being too rigid because current political, economic, social and cultural changes in Asian countries do not always dovetail with the concepts of 'urban' and 'rural'. The growth of cities in Asia, for instance, will not eliminate the huge rural population in these countries. According to Koppel (1991, pp.48-50), these special processes and their underlying causes cannot be explained by the urban-rural dichotomy.

The assumptions that underlie the urban-rural dichotomy, according to Koppel (1991, p.49), are confusing and misleading. The first puzzling premise is that rural society is dependent on its urban counterpart through a variety of interrelationships. Under this hypothesis, rural areas are regarded as regions responsible solely for agricultural production and isolated from urban markets and urban residents. Secondly, it is superficially assumed that the decline of agricultural activities and productivity is the result of an increase in urban economic influences. The third erroneous belief is that a particular set of political, economic, social and cultural relations is embedded in cities and the countryside. Therefore, the urban-rural dichotomy assumes that the same set of relationships can be found in different cities and their rural counterparts. Any studies based on these assumptions, and the urban-rural perspective, are said to be weakened by their failure to "recognise that the dichotomies are part of the problem" (Koppel, 1991, p.51). This is particularly the case in analysing the impacts of rural reforms and policies which favour agricultural production and its support system.

A different viewpoint from that of McGee, Koppel and Hawkins is presented by Potter and Unwin (1995). They focus on interactions between urban and rural areas in the developing world, which stem from recent urbanisation and global changes, and argue that these demonstrate the importance of the urban-rural perspective. Although the formation of functionally integrated zones has physically linked the city and the countryside, it is doubtful, according to Potter and Unwin (1995), whether rural areas will benefit from this integration during recessions and times of high international debt. They suggest that there are several conditions which are unfavourable to rural areas (Potter and Unwin, 1995, p.69). First, the concentration of private capital in the urban cores has grown, and as a result offshore manufacturing enterprises are drawn to the principal urban rather than rural areas. Secondly, adjustments to cope with recession may also be used to restrict the spread of new technology from urban areas. Finally, the exchange of agricultural and industrial products between urban and rural areas favours the former. This is because central places for economic activities – urban areas – always have a higher potential to generate revenue for both state and private investors. Consequently, "rather than growth spreading from urban to rural areas in these hard times, the rural zones seem to be paying an exceptionally high price for the polarised forms of urban development" (Potter and Unwin, 1995, p.69).

Moreover, global political changes, particularly the collapse of socialism in Eastern Europe and the former USSR, have boosted the spread of capitalism – a process which is intimately associated with urbanisation. Further, in order to incorporate developing countries into the world economic system – dominated by a market economy – and to attract foreign investment, they have had to expand their urban infrastructure. A similar situation is observed in China, where most foreign direct investment has been located in urban areas (Yeh, 2000). Consequently, the gap in development between urban and rural areas has widened.

The debate on urban-rural dichotomy is still ongoing. However, there does seem to be value in retaining the concept in order to understand the spatial development of Asian countries, particularly China. There is no doubt that rapid development in these countries, such as the emergence of rural industries, has blurred the distinction between urban and rural. However, the continuing juxtaposition of development and underdevelopment, prosperity and poverty, and rich and poor highlights a dilemma for McGee, Koppel and Hawkins. The outright rejection of the urban-rural perspective simply assumes that exploitation between regions (particularly those between city and countryside), between the agricultural and industrial sector, and unequal development between urban and rural areas no longer exists. In fact, in most Asian developing countries, the conflicts between formal and informal sectors, different policies in agricultural and non-agricultural activities, and the inequalities of income between rural and urban are problems too pronounced to be ignored. These conflicts and disparities are closely related, or produced, by the dichotomous development of these countries (Sethi, 1994, p.12). For example, Tang and Chung's (2000) study illustrates that economic integration between urban and rural areas in Jiangsu Province – one of China's most prosperous regions – has been constrained by the administrative system. Investigating illegal land use in an urban fringe, they further demonstrate an urban-rural fragmentation at a ground level (Tang and Chung, 2002). They argue that the unlawful behaviour of individual peasants is the result of China's upholding urban-rural segregation. This structure excludes peasants from the social welfare system. Coupled with the loss of cultivated land, to struggle for a living peasants have to extend their houses on the assigned residential land for leasing purposes. While spatial fragmentation and inequality still exists between sectors and regions, the idea of dismissing entirely the urban-rural perspective seems an over-reaction to the development of rural China.

The urban-rural notion still plays an important role in understanding China's spatial development during the reform period. This is because Chinese space, in Passi's (1991) terms, has been institutionalised into two hierarchies – urban and rural. According to him, the institutionalisation process is a dynamic one with regions continually being re-defined by social and economic processes (Passi, 1991). However, the dynamic of change has been frozen by China's political-cum-economic system. Since 1950, the implementation of the household registration system has divided Chinese society into urban and rural. During Mao's era, the former was administratively organised by cities and the latter by counties. Further, the city specialised in industrial production and the rural in agricultural activities. Thus, urban and rural areas were separated, with certain economic activities being organised and managed by different administrative bodies. Since inter-regional interactions were highly restricted, rural areas were regarded as closed economic units that specialised in food and raw material production. Economic activities in rural areas were determined by orders and plans sent down from the higher levels rather than by direct contacts with urban areas. Urban areas were closed in the same way.

Although economic reform since 1978 has broken down the rigid division of labour between urban and rural areas and has reactivated the interactions between them, the administrative structure has persisted (see Chapter 3). Thus, in the Chinese context, 'urban' and 'rural' cannot merely be seen as spatial concepts, they are also political notions that explain institutional and power differences over space. Concomitantly, urban-rural relations in China are not merely relationships of economic activities over space, but also between different administrative units. Accordingly, unlike Skinner's pure spatial core-periphery concept, the urban-rural notion is a geopolitical one which explains China's regional variations and provides "a link between policy and real power" (Sethi, 1994, p.17). This viewpoint remains a useful tool for understanding China's political economy and spatial development since 1949 in general, and the variations between Deqing and Dongguan in particular.

This new urban-rural concept helps scholars understand that variations between Deqing and Dongguan are not merely economic or spatial, but are also structural due to their administrative status. In the early 1980s, both Deqing and Dongguan were county-level units. However, after Dongguan was promoted to 'county-level city' and 'prefecture-level city' in 1985 and 1988 respectively, it has been granted an 'urban' status with associated powers and privileges. This has included greater economic autonomy, more power to approving land development projects and more central investments. Moreover, urban related authorities, such as departments of planning, power supply, transport and sewerage, have been established. In contrast, Deqing has persisted in its county position with little increase in power. Administratively, and statistically, it is still regarded as 'rural'.

Deqing and other rural districts have been further undermined by the city-leading-counties system. This is because counties are administratively subordinated to prefecture-level cities (leading cities) under the new system. Instead of helping the countryside to develop, the urban area exploits the resources of rural districts. Such a situation is regarded as 'city-extorting-counties' (*shi gua xian*) (Gao and Pu, 1987; Xia and Yang, 1991; Wang, 1992; Pu, 1995; Ma, 1996). Whether leading city Zhaoqing has extorted from Deqing is unknown. However, the system of city-leading-counties reinforces the unequal position between urban and rural areas. Also, it provides a mechanism for the unequal exchange of resources between urban and rural areas. Due, in part, to fewer central inputs, rural districts and agricultural activities are generally in subordinate positions. Disparities between urban and rural areas are expanding (URSE-Study Group, 1996a, 1996b; Fan and Xiang, 1999). Thus, variations in administrative status (i.e. urban and rural) have structurally determined Deqing and Dongguan's differences in economic and market development, not to mention the impact of preferential policies and disparities in foreign investment.

While case studies of Deqing and Dongguan illustrate variations in government regulations determining rural marketing activities, the urban-rural notion provides a useful framework for understanding how and why such variations have occurred. This goes beyond general descriptions of spatial differences by linking them to their political context and policies. It reminds researchers of the continuing importance of political factors in any study of China, despite economic reform weakening the capacity of the state to regulate. Moreover, the juxtaposition of government regulations and market forces in Deqing and Dongguan's market development demonstrates that a purely economic perspective is inadequate. While Huang (1996b) suggests political scientists adopt an economic perspective to understand local government behaviour, it is equally important for geographers to employ a geopolitical view in studying China's spatial development. This inter-disciplinary perspective is more attuned to China's complex reality during the era of economic reform.

The urban-rural concept is not a panacea for understanding all types of spatial variations in China. However, it is still a useful tool for gaining insights into China's regional development and its underlying determinants because economic, social and political evolutions are so different between urban and rural areas. While this study has highlighted gaps in market development between rural areas with different administrative status, other disparities between urban and rural market growth also require investigation. As noted in Chapter 3, the number of urban markets increased twice as fast as their rural counterparts between 1979 and 1997. Population growth does not provide a sufficient explanation for such expansion because there is an active birth control policy in the cities. Accordingly, why does this situation exist and how can it be explained? These questions require further research.

Apart from market development, a number of research questions need to be raised about the continuing legitimacy of urban-rural concepts. For example: (a) Why do economic reforms appear more successful in rural than in urban areas? (b) Why have democratic elections been launched in villages but not in cities? (c) Why has local corporatism emerged in the countryside while rarely being found in urban areas? and (d) Why are the rural 'new rich' so eager to pay money for urban household registrations? In considering these issues, it is necessary to look at how structural differences between urban and rural areas have created disparate forms and ways of development during the reform era. Classical geographers are in a good position to contribute to these discussions given their abiding interest in spatial variations.

To conclude, conceptualising spatial variations has never been an easy task in China because of its extraordinary scale. The case of Deqing and Dongguan shows neither Skinner's topographic models nor the core-periphery concept has provided a satisfactory explanation of China's rural marketing process. A major deficiency of these notions is their failure to provide a critical connection between spatial distribution and the context in which it is embedded. This emphasis is crucial to an understanding of differences and diversity, topics which rate highly in geographers' research agenda under the post-modern influence. Indeed, recent discussions on difference suggest it is a product of identity, which is socially and politically constructed. The perception has raised concerns about the importance of political process and power in interpreting spatial variation (Smith, 1999) – requiring examination of party affiliation, voting behaviour and social participation. The urban-rural notion in the Chinese context is an elaboration of this idea. It deals with the formal arena of a political structure by inviting investigations into China's administrative hierarchy and setting up a link between institutions, identity and difference.

The emphasis on embeddedness has shown a shift of focus from spatial pattern to process. This move is echoed by Cartier (2002), when she reviews Skinners' regional approach in general and marketing framework in particular. She suggests the regional boundary has lost its significance in defining places because the process of globalisation, the emergency of long-distance trade and the growth within regions has intensified transregional interactions (Cartier, 2002). As a result, investigation of marketing systems should go beyond morphological descriptions of spatial structure by looking at the progressive side of rural market networks. This includes its formation, evolution and the creation of local identity.

These changes have presented a new perspective on how geographers look at space. Under the influence of modernism, space was seen as a direct manifestation of economic processes. Studies, such as Skinner's rural marketing networks, were thus focussed on the mapping of economic attributes on space. The aim was to show the relations between certain socio-economic characteristics and particular spatial structures. Since the 1970s, this perspective has been challenged by Marxism and realist philosophies, which argue that spatial patterns both express and shape socio-economic relations. The concepts of spatial contingency, spatial boundary, and locality effects discussed in Chapter 1 are examples of this development. Recently, under the influence of postmodernism, scholars have suggested alternative perspectives to comprehend the role of space in explaining socio-economic activities. A notion of 'Third Space' is proposed as a third option to the binary conception of society and space (Soja, 1996). This perspective considers individuals' challenge and negotiation with the social worlds through their use of space during the construction of social and economic space. Thus, space is neither a reflection nor a causal factor of social and economic processes. It is regarded as a means of resistance and celebration. According to Smith (1999, p.20), "Thirdspace is created by those who reclaim these real and symbolic spaces of oppression, and make them into something else". These new emphases have pointed to the "complexity, ambiguity and multi-dimensionality of identity" of society (Tang and Lee, 2003, p.11). It offers not only a unique perspective to understand space and diversity, but also challenges the notion of sameness and grand theories, which attempt to generalise the human world.

Chapter 8

Skinner, Rural Market Development and Economic Reform: A Conclusion

Skinner's simple but inspirational framework on rural marketing was widely applied in developing countries during the 1970s (Bromley, 1971; Eighmy, 1972; Beals, 1975; Appleby, 1976; Plattner, 1976; Schwimmer, 1976; Symanski, 1978; Handwerker, 1978). None of these studies, however, found Skinner's models entirely satisfactory. This is because market improvement and its pattern are embedded in the historical, economic and social structures of a place. The re-examination of his thesis in China during the economic reform echoes this finding. Skinner's contention provides a well-developed framework for understanding China's rural market development and its relationship to the administrative system. Generally, his description of the association between market evolution and agricultural specialisation reflects the nature of China's rural market development since 1978. Factors addressed in his sophisticated framework, such as road routes, level of self-sufficiency and commercialisation, and government regulation, are still important factors shaping China's rural market development. Furthermore, Skinner's equation of improved road routes and rural market advancement proved to be valid in some areas of Deqing and Dongguan. Also, his predictions about the level of self-sufficiency and commercialisation during market modernisation almost precisely outline the general situation in the case-study areas.

Skinner's attempt to generalise rural market evolution, however, has oversimplified China's situation. While his hypotheses about distribution, intensification, modernisation of markets, and transport development have provided useful insights, his studies are not without problems.

Oversimplification has weakened Skinner's fundamental arguments. Although Skinner (1964, 1965a) used a number of case studies to illustrate his model of rural market modernisation, his description of rural districts near Chengdu, Shanghai and Ningbo were not truly representative of China's situation. These examples only reflect the local condition in a few parts of China, and merely provide a general picture of prosperous regions, notably the lower Yangzi region and the Sichuan basin. Similarly, Skinner's (1965a) use of two models to explain market patterns in plains and mountainous areas recognises the importance of topography but they fail to fully accommodate China's physical diversity. Since the connections between market development and the economic and social context are cut off, other differences created over time and space are neglected.

The separation of rural market change and its economic and social context has generated a sense of environmental determinism. The two models focus on how physical differences affect patterns of settlement and their economic and social development. Although Skinner has progressed beyond the homogeneous plain of Christaller and Lösch, his efforts still underplay geographical differences. The discipline of modern geography considers how space shapes and reflects a region's social, historical and cultural background. Consequently, concepts like spatial contingency effects, spatial boundary effects and locality effects have been introduced to tackle the problem. These concepts remind geographers of the significance of China's changing political economy and its impact on rural market growth. Moreover, they also highlight the importance of a study area's local economic, social, historical and cultural contexts and their interrelationships with market expansion. Instead of the plains and mountains models, the progress of China's marketing needs to be examined within distinct local contexts. This argument is particularly important given China's extraordinary geographical scale and diversity.

In particular, Skinner's over-generalised description has weakened the portability of his model beyond the Imperial period. Skinner (1985b) aligns China's policy changes with the country's rural market development using a policy cycle theory. However, he does not address structural forces such as the control mechanisms of a planned economy and the behaviour of local government. Accordingly, his study fails to comprehend why government intervention has persisted during the reform period, which he regarded as a 'liberal phase' (Skinner, 1985b).

Oversimplification has severed the connections between Skinner's theory and its social, economic and political contexts. Thus, the application of his models and concepts to Deqing and Dongguan provides an inadequate explanation of historical and geographical differences. The experiences of Deqing and Dongguan testify to the complexity of rural market development during the era of economic reform. This process is not merely a function of one or two key elements, but is generated by various social, economic, political and environmental factors, each of which has equal weight but is important in different ways. Moreover, a number of new components have emerged while old ones have changed their significance. As a result, while the role played by transport remains important, accessibility is not the only force contributing to market development. Foreign investment, rural industrialisation, environmental problems, government regulations, the level of self-sufficiency among peasants and their respective income levels are all significant in shaping China's rural market development during the reform era.

The case studies of Deqing and Dongguan demonstrate that no simple correlation exists between social, economic, political and environmental aspects and rural market growth. While transport elements are important generally in Dongguan and in some villages of Deqing, they are not the sole determinant elsewhere. Similarly, government regulations play an important role in influencing Deqing's rural marketing activities while their impact on Dongguan is very limited.

These variations highlight the close connection between market evolution and local contexts. Thus, economic, social and political factors have a different impact when they are embedded in different contexts. The causes and constraints on rural market advancement in Deqing are different from those in Dongguan and other places in China. In other words, rural market development has unique characteristics in every locale. These findings demonstrate that grand theories, which attempt to generalise the reality, are always inadequate in explaining differences between places.

The persistence of government regulation in Deqing reflects the significance of political factors in China. While economic reform has weakened the state's central control, that of local governments has been reinforced by the decentralisation of power. This has affected both central-local relations and the forms of government regulation. Indeed, local government intervention explains why Skinner's population-market equation and some hypotheses on market growth do not apply in Deqing and Dongguan. Thus, assessments on market development cannot readily be made by reference to Skinner's models because he does not address the structural forces and the changing political economy of China.

The inadequacy of Skinner's framework has not only challenged the legitimacy of grand theories, but also the application of Western hypotheses in China. Indeed, Skinner noticed the close relationship between administrative and economic hierarchies and the importance of the former in Chinese society. However, his analysis pinpointed economic components and argued that economic hierarchy was fundamental to an understanding of Chinese society (Skinner, 1977c). This argument stemmed from his attempt to use Western economic theories, based on rational choice, to explain a non-Western society. Such practice is criticised by Wang, Mingming (1997) as 'cultural imperialism' which overlooks cultural and historical differences between East and West. Nevertheless, whether social and economic theories based on Western experience can explain Chinese society is still debatable. While this work suggests a direction for rethinking the issue, much debate may occur before a consensus is reached.

This research has re-established connections between rural market development and its economic, social and political context at both national and local levels. While peasant-market interactions have been examined, connections between different local market systems have not been fully analysed here. Such an investigation would have to be on a much larger geographical scale and encompass scrutiny of flows of commodities and people between counties, thus suggesting a direction for future work.

This study can be amplified by looking at market growth in other areas and provinces. While Skinner (1994) identifies seven zones in the Lingnan (Guangdong and Guangxi) region, it is imperative to study different levels of market improvement within these zones. In doing so, one would need to appreciate the spatial variations of rural market advancement and their underlying causes. Also it would be worthwhile to compare rural market systems in different macroregions to provide a more complete picture of the role of spatial differentiation in rural market development in post-reform China.

Invisible hand or invisible wall: Prospects for rural market development and China's reform

The joint functions of the 'invisible wall' and the 'invisible hand' in Deqing and Dongguan suggest that a hybrid market system has been formed. This structure is a unique product of China's economic reform, and its emergence challenges the assumption that market forces and central planning cannot co-exist in a functional economy (Muldavin, 1998). Also, this scheme implies that economic reform since 1978 has yet to find an ideal path in China. Accordingly, old and new systems, market forces and government intervention continue to co-exist.

Whether a pure market-driven system – Skinner's 'natural' system – will emerge is yet to be seen. While more emphasis has been put on the invisible hand during the reform era, the importance of government regulation in China's rural market development remains intact. Li, Maowen (1999) praises the establishment of an Office of Market Establishment (*shichang jianshe bangongshi*) by the city government and the practice of centralised planning in establishing rural markets in Henan Province. He suggests administrative measures are necessary to regulate rural market development. Peng (1999) goes further in arguing that the state should play an active role in establishing intermediate agents to improve China's rudimentary rural market structure, particularly in mountainous areas. Moreover, Li, Bingkun (1999) suggests that the state has to play a more important role in establishing a modern market network with a centralised core of wholesale markets connecting rural market and urban retail systems.

Spatial variations in rural market development between coastal and inland areas and prosperous and poor hilly areas have also inspired calls for greater state regulation among Chinese scholars. They argue that the central government needs to have functions which direct, support, protect, standardise, regulate and monitor rural market growth under a socialist market economy. One of the major tasks of government regulation, according to Sheng and Chen (1996), is to maintain 'balanced growth' in the rural market system between different regions. This is crucial for the establishment of a rational distribution mechanism which links central and local wholesale markets, rural market-places and urban retail systems. Also, Jiang (1999) urges local governments to put more effort into developing their sphere of circulation to help rural markets flourish.

It is not surprising to find that many Chinese scholars strongly emphasise the need for government regulation. In fact, unequal growth between urban and rural areas, irrational location and distribution of rural markets, and poor connections between some regional wholesale and local markets have aroused dissatisfaction with the country's market development. These factors entail significant drawbacks to the reform of China's rural economy which involves 70 per cent of the country's population. This is particularly true when the state has decided to stimulate rural market growth so as to maintain a minimal national economic growth of 8 per cent annually.

The state ICMB has recognised the above requests. Indeed, Cao (1996) has stressed the importance of regulating rural market improvement. The Bureau insists that market growth must follow the principle of 'state control, central planning, local government construction of markets, rational distribution of markets and ICMB monitoring'. This principle does not have any legal effect but it provides a powerful mechanism for government intervention. Nevertheless, it is difficult to guarantee that central policies on market development will be implemented at the local level, given the complex situation in the reform era and the changing relationship between central and local governments. This is because local governments now have enormous power within their jurisdictions. Whether or not rural markets are well built is a direct result of local government intervention, not economic forces.

The call for state intervention demonstrates the shortcomings of market forces in shaping China's rural market development. However, some defects are generated by government regulation. Accordingly, the request for regulation to correct errors made by government intervention is a strange logic. This muddled thinking is not surprising because China has been regarded as a country "riddled with policy contradiction" during the reform era (Oi, 1999a, p.627). Attempts in Deqing to use administrative measures to guide rural market development have been problematic. Apparently, China's rural market evolution has lost its direction on the way towards a socialist market economy.

Obviously, the paradox is generated by a double track system which permits the co-existence of central planning and market forces, and intervention and free competition. The old system based on central planning remains the core while the new one – hinged on market forces – has developed around it. This juxtaposition of the old and new has created serious problems. A significant consequence has been the intensification of 'rent seeking' behaviour that generates bribery and corruption, drains state resources and distorts the economy.[1] This is because the old and new systems, symbolised by state authorities and market forces respectively, cannot be reconciled. Once the authorities created a privileged 'rent seeking' niche, market forces were distorted and a level playing field was difficult to achieve (Fan and Gao, 1998). Moreover, gaps occurred during the transition period when the new system was being established, intensifying the conflict between the rival systems. Whether China will successfully make a transition to a market economy

---

[1] The concept of 'rent seeking' has been widely applied to studies of local government behaviour. It refers to "behaviour in institutional settings where individual efforts [are made] to maximise value [of personal interest]" (Buchanan, 1980, p.4). See Lu's (1994) study for an application of this concept to China's private enterprises development.

depends on the growth of the new system and the support of a well-established legal scheme.

The double track structure is a unique product of China's gradual reform process – a great leap economically but a tiny step politically. As political and social stability has been the first priority in this gradualist approach, China has successfully avoided large-scale social and political turmoil during the past two decades. However, many sensitive areas remain untouched by reform for the same reason. In other words, China's gradual reform protects those with vested interests in the old system (Min, 1995). Many structural problems persist and examples of these are plentiful. For instance, China's market liberalisation proceeds "on the condition that a steady supply of low-cost food" is available (Rozelle, 1996, p.198). This is a political rather than an economic consideration because social and political instability will arise if the state fails to feed its 1.3 billion population. The trade-off of this approach is a slow increase in the income of peasants, and stagnant growth of rural marketing activities. Consequently, agricultural production has generally been overlooked, agricultural products have been 'under-priced' and the incentive for peasants to perform agricultural activities has been dampened.

Thus, the efficiency of the gradualist approach to reform has been minimal in the countryside and the cost is high. Despite this, China's reform process, including rural market development, marketisation of grain, cotton and other important agricultural products, will persist on the basis of a gradual approach. This is because the strategy has brought about considerable improvements in both industrial and agricultural productivity without causing any chronic economic problems (Gang, Perkins and Sabin, 1999). Socially, many individuals and groups have benefited from the reform despite the growing gap between rich and poor. These people do not desire any radical changes in both economic and political aspects. In this way, a stable social and political environment for further reform has been preserved.

There is little political pressure for China to undertake radical changes. Since the economic reform to fast-track China's development was initiated by the state, the regime is unlikely to lose its legitimacy. Indeed, unlike Eastern European countries and the former Soviet Union, where dramatic political changes have occurred, China remains politically stable. Accordingly, a radical approach is very unlikely to be adopted in China and the country will continue to follow its gradualist strategy. This also implies that political reform will be slow. As long as the double track structure persists, the hybrid market scheme will become an end state. In this case, China's reform "is not [making a transition] from communism to capitalism, but from central planning to a hybrid system" (Muldavin, 1998, p.290). The rural market, as an arena where market forces and state control compete, remains under the contradictory influence of both the invisible hand and the invisible wall.

Appendix. Rural markets in Deqing county and their general situations, 1997

	Decheng Central Market	***Decheng Western Market***	***Decheng Eastern Market***	***Datang Rural Market***	***Guanxu Rural Market***	***Maxu Rural Market***
Location	Decheng town (the county seat)	Decheng town (the county seat)	Decheng town (the county seat)	Huilong town, Datang Administrative Zone	Guanxu town, town government seat	Maxu town, town government seat
Total area (m^2)	7266.25	2530.96	2279.76	344	1113.21	1128.9
No. of people visited (in millions/year)	2.92	1.09	1.83	0.29	0.36	0.44
Market days	2, 5, 8, 10	2, 5, 8, 10	2, 5, 8, 10	3, 6, 9	3, 6, 9	1, 4, 7
No. of fixed stalls	506	101	108	41	110	118
Open days	1992, December	1992, September	1988, May	1987, May	1990, May	1995, January
Established by*	ICMB	ICMB	ICMB	ICMB	ICMB	ICMB
Market type**	C	C	C	C	C	C
Annual turn-over value (in million *yuan*)	197.3	52.6	51.2	2.6	115.7	127.5

Note: * ICMB denotes Industry and Commerce Management Bureau; * * C denotes comprehensive market.
Source: Interview of ICMB officials, 1998.

Appendix. Rural markets in Deqing county and their general situations, 1997 (*continued*)

	Maxu Farm-cattle Market	***Gaoliang Rural Market***	***Mocun Rural Market***	***Yongfeng Rural Market***	***Yongfeng Farm-cattle Market***	***Bozhi Rural Market***
Location	Maxu town, town government seat	Gaoliang town, town government seat	Mocun town, town government seat	Yongfeng town, town government seat	Yongfeng town, town government seat	Bozhi town, town government seat
Total area (m^2)	n.a.	2530.96	2279.76	344	1113.21	1128.9
No. of people visited (in millions/year)	0.07	0.55	0.47	0.46	0.15	0.47
Market days	1, 4, 7	2, 5, 8	1, 4, 7	2, 5, 8	2, 5, 8	3, 6, 9
No. of fixed stalls	20	201	68	95	40	108
Open days	n.a.	1997, September	1992, May	1992, May	1983, April	1987, September
Established by	ICMB	ICMB	ICMB	ICMB	ICMB	ICMB
Market type	S*	C	C	C	S	C
Annual turn-over value (in million *yuan*)	2.7	132.7	9.9	18.25	7.97	131.1

Note: * S denotes specialised markets.

Appendix. Rural markets in Deqing county and their general situations, 1997 (*continued*)

	Wulong Rural Market	***Fengcun Rural Market***	***Jiushi Rural Market***	***Liucun Rural Market***	***Yuecheng Rural Market***	***Shapang Rural Market***
Location	Wulong town, town government seat	Fengcun town, town government seat	Jiushi town, town government seat	Jiushi town, Liucun Administrative Zone	Yuecheng town, town government seat	Shapang town, town government seat
Total area (m^2)	786.44	5466.7	1725	528	5616.81	230
No. of people visited (in millions/year)	0.29	0.49	0.45	0.22	0.47	0.18
Market days	1, 4, 7	1, 4, 7	3, 6, 9	2, 5, 8	2, 5, 8	5, 10
No. of fixed stalls	54	111	111	48	146	40
Open days	1988, May	1989, May	1993, August	1974, April	1991, May	1990, June
Established by	ICMB	ICMB	ICMB	ICMB	ICMB	ICMB
Market type	C	C	C	C	C	C
Annual turn-over value (in million *yuan*)	8	13.5	12.4	6	13	5

Bibliography

Agricultural Ministry (1995) "Wuguo nongchengpin peifashichang jianshe yinjiu (Study on China's agricultural wholesale market", *Zhongguo Nongcun Jingji (Chinese Rural Economy)*, No.10, pp.10-19.

Appleby, G. (1976) "The role of urban food needs in regional development, Puno, Peru", in Smith, R. T. H. (eds) *Market-Place Trade – Periodic Markets, Hawkers and Traders in Africa, Asia, and Latin America*. Vancouver: University of British Columbia, pp.147-178.

Ash, R. F. (1996) "The peasant and the state", in Hook, B. (ed.) *The Individual and the State in China*. New York: Oxford University Press, pp.70-97.

Beals, R. L. (1975) *The Peasant Marketing System of Oaxaca, Mexico*. Berkeley: University of California Press.

Berry, B. J. L. (1959) "Recent studies concerning the role of transportation in the space economy", *Annuals of the Association of American Geographers*, Vol.49, pp.328-342.

Berry, B. J. L. (1967) *Geography of Market Centers and Retail Distribution*. Englewood Cliffs: Prentice Hall.

Berry, B. J. L. and Horton, F. E. (1970) *Geographic Perspectives on Urban Systems*. Englewood Cliffs: Prentice Hall.

Blecher, M. and Shue, V. (1996) *Tethered Deer: Government and Economy in a Chinese County*. Stanford: Stanford University Press.

Bromley, R. J. (1971) "Markets in the developing countries: A review", *Geography*, Vol.56, No.2, pp.124-132.

Bromley, R. J. (1978) "Traditional and modern change in the growth of systems of market centres in highland Ecuador", in Smith, R. T. H. (ed.) *Market-Place Trade – Periodic Markets, Hawkers and Traders in Africa, Asia, and Latin America*. Vancouver: University of British Columbia, pp.31-47.

Brown, G. P. (1998) "Budget, cadres and local state capacity in rural Jiangsu", in Christiansen, F. and Zhang, Junzuo (eds) *Village Inc.: Chinese Rural society in the 1990s*. United Kingdom: Curzon, pp.22-47.

Bryan, J., Hill, S., Munday, M. and Roberts, A. (1997) "Road infrastructure and economic development in the periphery", *Journal of Transport Geography*, Vol.5, pp.227-237.

Buchanan, J. M. (1980) "Rent seeking and profit seeking", in Buchanan, J. M., Tollison, R. D. and Tullock, G. (eds) *Toward a Theory of the Rent-seeking Society*. USA: Texas A & M University Press, pp.3-15.

Cao, Tiandian (1996) "Gaige zhong de zhongguo nongcun shichang (China's rural market place in reform)", *Jingji Yanjiu Cankou (Reference on Economic Studies)*, No.43, pp.2-10.

Cartier, C. (2002) "Origins and evolution of a geographical idea: The macroregion in China", *Modern China*, Vol.28, pp.79-142.

CASS-RDI and SSB-RSESG (Chinese Academy of Social Science Rural Development Institute and State Statistical Bureau Rural Social Economic Survey Group) (1998) *1997-1998 Nian: Zhongguo Nongcun Jingjixinshi Fenxi Yu Yuce*

(Analysing and Forecasting China's Rural Economic Situation, 1997 to 1998). Beijing: Shehui Kexue Wenxian Chubanshe.

Chan, K. W. (1992) "Economic growth strategy and urbanisation policies in China, 1949-1982", *International Journal of Urban and Regional Research*, Vol.16, pp.134-50.

Chen, Qiang (1997) "1997-1998 nian wuguo nongcun shichang fazhen qianjian fenzhi (An analysis on China's rural market development during 1997-1998)", in Guojia Xinxi Zhongxin (State Information Centre) (eds) *Zhongguo Shichang Jianwang (Outlook on China's Market)*. Beijing: Zhongguo Jihua Chubanshe, pp.103-108.

Chiu, T. N. and Leung, C.K. (1983) "Periodic markets and spatial organization in Gaohe", in Leung, C.K. and Chin, S.S.K. (eds) *China in Readjustment*. Hong Kong: University of Hong Kong Press, pp.247-261.

Christaller, W. (1966) *Central Places in Southern Germany*. Translated by Baskin, W. C., Englewood Cliffs: Pretice Hall.

Chung, H. (1994) *Geographic Transfer of Resources Under the Institutional Reform of City-Leading-Counties: With Special Reference to The Sunan Area*. Unpublished M. Phil. Thesis. Department of Geography, The Chinese University of Hong Kong.

Chung, H. (2002) "Some socio-economic impacts of toll roads in rural China", *Journal of Transport Geography*. Vol.10, pp.145-156.

Chung, H. and Tang, W.-S. (1996) "The 'city-leading counties' system reform and China's urban-rural relations", in Li, S. M. and Tang, W.-S. (eds) *Perspectives on China's Regional Economy*. Taipei: Population Institute, pp.307-333. (In Chinese).

Chung, H. and Tang, W.-S. (1997) *Regionalism under Deng: Localism Centred around Cities and Towns*. Paper presented at the 5th Conference of the Chinese Studies Association of Australia, University of Adelaide.

Chung, J. H. (1995) "Studies of central-provincial relations in the People's Republic of China: A mid-term appraisal", *China Quarterly*, No.142, pp.487-508.

Cong, Hanxiang (1995) *Jindai Lu Yu Xiangcun (Hebei, Shandong, Henan Villages during Modern China)*. Beijing: Zhongguo Shihui Kexue Chubanshe.

Cox, K. R. (1995) "Concepts of space, understanding in human geography, and spatial analysis", *Urban Geography*, Vol.16, pp.304-326.

Crissman, L. W. (1976a) "Specific central-place models foe an evolving system of market town on the Changhua plain, Taiwan", in Smith, C. A. (ed.) *Regional Analysis. Volume One: Economic System*. London: Academic Press, pp.183-218.

Crissman, L. W. (1976b) "Spatial aspects of marriage patterns as influenced by marketing behaviour in West Central Taiwan", in Smith, C. A. (ed.) *Regional Analysis. Volume Two: Social System*. London: Academic Press, pp.123-148.

Deqing County Gazetteer Editorial Committee (1996) *Deqing Xianzhi (Deqing County Gazetteer)*. Guangdong: Renmin Chubanshe.

DGSB (Dongguan Statistical Bureau) (1995) *Dongguan Tongji Nianjian 1995 (Dongguan Statistical Yearbook, 1995)*. Unpublished material.

DGSB (Dongguan Statistical Bureau) (1998) *Dongguan Tongji Nianjian 1998 (Dongguan Statistical Yearbook, 1998)*. Beijing: Zhongguo Tongji Chubanshe.

DGSB (Dongguan Statistical Bureau) (1999) *Dongguan Tongji Nianjian 1999 (Dongguan Statistical Yearbook, 1999)*. Beijing: Zhongguo Tongji Chubanshe.

DGSB (Dongguan Statistical Bureau) (2001) *Dongguan Tongji Nianjian 2001 (Dongguan Statistical Yearbook, 2001)*. Beijing: Zhongguo Tongji Chubanshe.

Ding, Shenjin (1998) "Peiyu he wanshan nongcun shichang tixi (Cultivate and enhance rural market system", in Wang, Hei and Duan, Chihuang (eds) *Shichang Jingji Yu Zhongguo Nongye Wanti Yu Qianjing (Market Economy and China's Agriculture: Problems and Prospects)*. Shanghai: Fudan Daixue Chubanshe, pp. 40-55.

DQSB (Deqing Statistical Bureau) (1996) *Deqing Tongji 1949-1995 Nian (Deqing Statistics: 1949-1995)*. Unpublished material.

DQSB (Deqing Statistical Bureau) (1998) *Deqing Xian Guominjingji Tongji Ziliao Huibian 1997 (A Compliation of Deqing's National Economic Statistic, 1997)*. Unpublished material.

DQSB (Deqing Statistical Bureau) (1999) *Deqing Xian Guominjingji Tongji Ziliao 1949-1998 (National Economy Statistic of Deqing County, 1949-1998)*. Unpublished material.

DQSB (Deqing Statistical Bureau) (2002) *Deqing Xian Guominjingji Tongji Ziliao 2000-2001 (National Economy Statistic of Deqing County, 2000-2001)*. Unpublished material.

DTM (Domestic Trade Ministry of People's Republic of China) (2001) *Zhongguo Guonei Maoyi Nianjian 2001 (Almanac of China's Domestic Trade 2001)*. Beijing: Zhongguo Guonei Maoyi Nianjian Chubanshe.

Du, Zuofeng (2001) "Nongcun shichangwangluo de wanshan yu chengshihua de tuijin (Improving rural marketing networks and promoting urbanisation)", *Zhongguo Nongcun Jingji (Chinese Rural Economy)*, No.9, pp.10-15.

Duncan, S. (1989a) "Uneven development and the difference that space makes", *Geoforum*, Vol.20, pp.131-139.

Duncan, S. (1989b) "What is locality?", in Peet, R. and Thrift, N. (eds) *New Models in Geography: the Political-Economy Perspective*. London: Unwin Hyman, pp.221-252.

Edin, M. (1998) "Why do Chinese local cadres promote growth? Institutional incentives and constraints of local cadres", *Forum for Development Studies*, No.1, pp.97-127.

Eighmy, T. H. (1972) "Rural periodic markets and the extension of an urban system: A Western Nigeria example", *Economic Geography*, Vol.28, pp.299-315.

Elvin, M. and Skinner, G. W. (eds) (1974) *The Chinese Cities Between Two Worlds*. Stanford: Stanford University Press.

Eng, I. (1997) "The rise of manufacturing towns: Externally driven industrialisation and urban development in the Pearl River Delta of China", *International Journal of Urban and Regional Research*, Vol.21, pp.554-568.

Fan, C. C. (1997) "Uneven development and beyond: Regional development theory in Post-Mao China", *International Journal of Regional Studies*, Vol.21, pp.620-639.

Fan, Gang and Gao, Minghua (1998) "Zhuangui guodu: Zhongguo shichang tixi de xingcheng he fazhan (Double track transition: The formation and development of China's market system)", in Hang, Zhiguo, Fan, Gang, Liu, Wei and Li, Yang (eds.), *Zhongguo Gaige yu Fazhan de Zhidu Xiaoying (Institutional Impact on China's Reform and Development), Vol. 1*. Beijing: Jingji Kaxue Chubanshe, pp.102-140.

Fan, Jianping (1996) "Shixian xiaokang mubiao de liandan yu duice (The difficulties and solutions to achieve well-off status)", *Jingji Yanjiu Cankou (Reference on Economic Studies)*, No. 22/23, pp.2-34.

Fan, Jianping and Xiang, Shujian (1999) "Wuguo chengxiang renkou eryuan shehui jiegou duijumin xiaofeilu de yingxiang (The dual structure of urban-rural population and its impact on consumption)", *Guanli Shijie (Management World)*, No.5, pp.35-63.

Fei, Hsiao-Tung (1953) *China's Gentry: Essays on Rural-Urban Relations*. Chicago: The University of Chicago Press.

Findlay, C., Watson, A. and Martin, W. (1993) *Policy Reform, Economic Growth and China's Agriculture*. Paris: OECD.

Forster, K. (1991) *China's Tea War*. Working Paper No. 91/3. Chinese Economy Research Unit. The University of Adelaide.

Fulton, D. C. (1969) "A road to the West", *Finance and Development*, Vol. 6, No. 3, pp.2-7.

Funnell, D. C. (1988) "Urban-rural linkage: Research themes and directions", *Geografiska Annaler*, Vol.70, Series B, pp.267-274.

Gang, F., Perkins, D. H. and Sabin, L. (1999) "People's Republic of China: Economic performance and prospects", *Asian Development Review*, Vol.15, pp.43-85.

Gao, Wangling (1985) "Qian-jia shiqi Sichuan de shichang he nongcun jingji jiegou (Sichuan's rural market and rural economic structure during the qian-jia period)", in Zhongguo Nongcun Fazhan Wanti Yanjiuzu (eds) *Nongcun, Jingji, Shihui, Di yi juan (Rural, Economy, Society, Vol.1)*, Publisher missing, pp. 338-352.

Gao, Yan and Pu, Shanxin (1987) "Lun shiguanxian tizhi (Discussions on the system of city-leading-counties)", *Zhongguo Chengzhen (Chinese City and Township)*, No.4, pp.36-39.

Garnaut, R. and Ma, G. (1996) "The third revolution", in Garnaut, R. and Ma, Guonan (eds) *The Third Revolution in the Chinese Countryside*. United Kingdom: Cambridge University Press, pp.1-9.

Garrison, W. L., Berry, B. J. L., Marble, D. F., Nystuen, J. D. and Morrill, R. L. (1959) *Studies of Highway Development and Geographic Change*. Seattle: University of Washington Press.

GDPN and GDICMB (Guangdong Committee of Place Names and Guangdong Industry and Commerce Management Bureau) (eds) (1992) *Guangdong Xuji (Market Places in Guangdong)*. Guangdong: Ditu Chubanshe.

GDSB (Guangdong Province Statistic Bureau) (1998) *Gangdong Tongji Nianjian 1998 (Guangdong Statistical Yearbook 1998)*. Beijing: Zhongguo Tongji Chubanshe.

GDSB (Guangdong Province Statistic Bureau) (1999) *Gangdong Tongji Nianjian 1999 (Guangdong Statistical Yearbook 1999)*. Beijing: Zhongguo Tongji Chubanshe.

GDSB (Guangdong Province Statistic Bureau) (2001) *Gangdong Tongji Nianjian 2001 (Guangdong Statistical Yearbook 2001)*. Beijing: Zhongguo Tongji Chubanshe.

Gong, Fentao, Kang, Changjin, Su, Zhongping and Wu, Xiyun (2001) "Dangqian Nongchanpinshichang de jibenzhuangkuang ji cunzai de zhuyao wenti (Basic situation and major existing problems of rural commodity markets)", *Zhongguo Nongcun Jingji (Chinese Rural Economy)*, No.2, pp.68-72.

Gong, Gang (1998) "Kaitou nongcun shichang wenti yingjiu (A study on the problems of opening rural markets)", *Zhongguo Liutong Jingji (China Business and Market)*, No.5, pp.31-34.

Gong, Jitong (1997) "Nongcun shichang de xiangzhuan, wenti he duici jianyi (Current situation, problems and suggestions on rural market development)", *Jingji Yanjiu Cenkou (References on Economic Studies)*, No.93, pp.16-23.

Good, C.M. (1973) "Market in Africa: A review of research themes and the question of market origins", *Cahiers d'Etudes Africaines*, Vol.13, pp.769-780.

Green, F. H. W. (1958) "Notes on the hierarchy of central places and their hinterlands", *Economic Geography*, Vol.34, pp.210-226.

Gregor, A. J. (1995) *Marxism, China, and Development: Reflections on Theory and Reality*. New Brunswick: Transaction Publishers.

Gu, Shengzu and Li, Zhen (1995) *The Impact of Reform on Structural Change in the Chinese Economy: The Case of Hubei Province*. Working Paper on East Asian Economic Studies. No.24. University of Duisburg. Germany.

Guo, Shutian and Liu, Chunbin (1990) *Shiheng De Zhongguo (The Unbalanced China)*. Heibei: Renmen Chubanshe.

Hajj, H. and Pendakur, V.S. (2000) *Roads Improvement for Poverty Alleviation in China*. Washington: The World Bank.

Han, Jin (1998) "Kaitou nongcun shichang zhonghan tian (Discussions on China's rural market development)", *Zhongguo Nongcun Jingji (Chinese Rural Economy)*, No.5, pp.4-6.

Han, Jin and Yu, Xian (1995) "1994-1995 Zhongguo nongcun jingji xingshi huigu yu zhanwan (Reviewing and projection of China's rural economy, 1994-1995)", *Zhongguo Nongcun Jingji (Chinese Rural Economy)*, No.2, pp.11-20.

Han, Jin and Yu, Xian (1996) "1995-1996 Zhongguo nongcun jingji xingshi huigu yu zhanwan (Reviewing and projection of China's rural economy, 1995-1996)", *Zhongguo Nongcun Jingji (Chinese Rural Economy)*, No.2, pp.3-12.

Handwerker, W. P. (1978) "Viability, location and timing of Liberian periodic markets", in Smith, R. T. H. (ed.) *Market-Place Trade – Periodic Markets, Hawkers and Traders in Africa, Asia, and Latin America*. Vancouver: University of British Columbia, pp. 199-221.

Ho, S. P. S. (1994) *Rural China in Transition: Non-agricultural Development in Rural Jiangsu, 1978-1990*. Oxford: Clarendon Press.

Hodder, R. (1993) *The Creation of Wealth in China: Domestic Trade and Material progress in a Communist State*. New York and London: Belhaven Press.

Hu, Angang (1999) *Zhongguo Fazhen Qianjian (The Prospect of China's Development)*. Zhejiang: Renmin Chubanshe.

Hu, Dayuan (1998) "Zhuangui jingji zhong de diqu chaju (Regional disparities in a transitional economy)", *Zhanlue yu Guanli (Strategy and Management)*, No.1, pp.35-41.

Hua, Meijia (2001) "Zhongguo nongchenpin paifashichang de jianshe yu fazhenfangxiang (Constructions and development trends of China's agricultural wholesale markets)", *Zhongguo Nongcun Jingji (Chinese Rural Economy)*, No.12, pp.37-41.

Huang, Guoqin and Wang, Xiaoyong (1993) "Nongfuchanpin shichang zenmouli (What happened to the markets for agricultural sidelines?)", *Shichang yu Fazhan (Market and Development)*, No.2-3, pp.25-26.

Huang, Jikun and Rozelle, S. (1998) "Market development and food demand in rural China", *China Economic Review*, Vol.9, pp.25-45.

Huang, Mingda (1993) *Chengzhentixi De Jiegou Yu Yanhua: Lilun Fenxi Yu Shizheng Yinjiu* (The Structure and Changes of Urban-Township System: Theoretic Analysis and Empirical Studies), Unpublished PhD. Thesis. East China Normal University. Shanghai.

Huang, Y. S. (1996a) "Central-local relations in China during reform era: The economic and institutional dimensions", *World Development*, Vol.24, pp.655-672.

Huang, Y. S. (1996b) *Inflation and Investment Controls in China: The Political Economy of Central-Local Relations during the Reform Era*. Cambridge: Cambridge University Press.

Huang, Yiping (1998) *Agricultural Reform in China*. United Kingdom: Cambridge University Press.

Jia, Daming (1999) "Nongmin zengshou shiguan quanju (Increasing the income of peasants is critical to China's whole economic situation)", *Gaige Neicen (Inside Information on Economic Reform)*, Vol.17, pp.13-15.

Jiang, Chengzhou and Liu, Weidong (2000) "Ruhe qidong kaituo nongcun shichang xuqiu (How to initiate and develop rural market demands", *Zhongguo Nongcun Jingji (Chinese Rural Economy)*, No.7, pp.52-54.

Jiang, H. (1993) "Distributions of rural centers near Chengdu in Southwest China", *Erdkunde*, Vol.47, pp.212-218.

Jiang, Jiyu (1999) "Kaitou nongcun shichang de zhiyue yinsu fenshi ji duice jianyi (An analysis of underlying constrains on rural market development and suggestions)", *Nongye Jingji Wanti (Problems on Rural Economy)*, No. 7, pp.43-46.

Jiang, Shanhe and Hall, R. H. (1996) "Local corporatism and rural enterprises in China's reform", *Organisation Studies*, Vol.17, pp.929-952.

Johnson, G. E. (1994) "Open for business, open to the world: Consequences of global incorporation in Guangdong and the Pearl River Delta", in Lyons, Thomas P and Nee, Victor (eds) *The Economic Transformation of South China: Reform & Development in the Post-Mao Era*. New York: Cornell East Asia Program, pp.55-87.

Kirkby, R. J. R. (1985) *Urbanisation in China: Town and Country in a Developing Economy 1949-2000 AD*. London: Croom Helm Ltd.

Knapp, R. G. (1971) "Marketing and social patterns in rural Taiwan", *Annals of the Association of American Geographers*, Vol. 61, pp.131-155.

Koppel, B. (1991) "The rural-urban dichotomy reexamined: Beyond the ersatz debate", in Ginsburg, N., Koppel, B. and McGee, T.G. (eds) *The Extended Metropolis: Settlement Transition in Asia*. Honolulu: University of Hawaii Press, pp.47-70.

Koppel, B. and Hawkins, J. (1994) "Rural transformation and the future of work in rural Asia", in Koppel, B. Hawkins, J. and James, W. (eds) *Development or Deterioration? Work in Rural Asia*. London: Lynne Rienner Publishers, Inc, pp.1-46.

Kornai, J. (1980a) *Economic of Shortage. Vol. A*. Amsterdam: North-Holland.

Kornai, J. (1980b) *Economic of Shortage. Vol. B*. Amsterdam: North-Holland.

Lado, C. (1988) "Some aspects of rural marketing systems and peasants farming in Maridi district, Southern Sudan", *Transactions Institute of British Geographers*, Vol.13, pp.361-374.

Lau, P. K. (1998) "Industry and trading", in Yeung, Yue-Man and Chu, D. K. Y. (eds) *Guangdong: Survey of a Province Undergoing Rapid Change (second edition)*. Hong Kong: The Chinese University of Hong Kong Press, pp.127-149.

Lee, P. K. (1998) "Local economic protectionism in China's economic reform", *Development Policy Review*, Vol.16, pp.281-303.

Lees, Francis A. (1997) *China Superpower: Requisites for High Growth*. New York: St. Martin's Press.

Leinbach, T. R. and Chia, L. S. (1989) "The role of transport in development", in Leinbach, T. R. and Chia, L. S., *South East Asian Transport: Issue and Development*. New York: Oxford University Press, pp.1-7.

Li, Bingkun (1999) "Nongchanping liutong tizhi gaige yu shichang zhidu jianshe (The reform of the agricultural product circulation system and market system construction)", *Zhongguo Nongcun Jingji (Chinese Rural Economy)*, No. 6, pp.11-18.

Li, Hua (1989) "Ming qing Guangdong xushi yanjiu (A study of Guangdong's rural market during the Ming and Qing Dynasty)", in Pinghuai xuekan editorial committee, *Pinghuai Xue Kan: Zhongguo Shihui Jingji Yanjiu Lunji (Papers on China's Socio-economic History, Vol.4 part II)*, pp.311-362.

Li, Maowen (1999) "Nongcun shichang jianshezhong cunzai de wanti ji shencengci yuanyin fenshi (An analysis of the problems and underlying causes for rural market development)", *Nongye Jingji Wanti (Problems on Rural Economy)*, No. 1, pp.45-47.

Li, S. M. (1996) "China's changing regional disparities: A review of empirical studies", in Li, S. M., Tang, W.-S., Jiang, Lanhong and Zhou, Suqing (eds) *Perspectives on China's Regional Economy*. Taipei: Population Institute, pp.19-43. (In Chinese).

Li, Wenyan (1990) "Contemporary spatial issues", in Linge, G. J. R. and Forbes, D. K. (eds) *China's Spatial Economy: Recent Developments and Reforms*. Hong Kong: Oxford University Press, pp.59-84.

Li, Zehua (2002) "Wuguo nongchanpin paifashichang de xianzhuang yu fazhenqushi (The current situation and development trends of China's agricultural wholesale markets)", *Zhongguo Nongcun Jingji (Chinese Rural Economy)*, No.6, pp.36-42.

Liao, Jili, Gao, Yishen and Zhou, Shujun (1991) *Lun Jihua Jingji (A Discussion on the Planned Economy)*. Guangdong: Gaodeng Jiaoyu Chubanshe.

Lin, G. C. S. (1993) "Small town development in socialist China: A functional analysis", *Geoforum*, Vol.24, pp.327-338.

Lin, G. C. S. (1997a) "Development and planning of small towns in China: Speculation, reassessment and prospect", *China City Planning Review*, Vol.13, pp.24-30.

Lin, G. C. S. (1997b) "Transformation of a rural economy in the Zhujiang Delta", *The China Quarterly*, No.149, pp.56-80.

Lin, G. C. S. (1997c) *Red Capitalism in South China: Growth and Development of the Pearl River Delta*. Vancouver: UBC Press.

Lin, G. C. S. (2001a) "Evolving spatial form of urban-rural interaction in the Pearl river Delta, China", *Professional Geographer*, Vol. 53, pp.56-70.

Lin, G. C. S. (2001b) "Metropolitan development in a transitional socialist economy: Spatial restructuring in the Pearl River Delta, China", *Urban Studies*, Vol.38, pp.383-406.

Lin, G. C. S. and Ma, L. J. C. (1994) "The role of towns in Chinese regional development: The case of Guangdong province", *International Regional Science Review*, Vol.17, pp.75-97.

Liu, Junde (2002) "Study on the innovation in the administrative organization and management of the metropolitan areas in mainland China, with special reference to the Pearl River Delta", in Wong, K.Y. and Shen, Jianfa (eds) *Resource Management, urbanization and Governance in Hong Kong and the Zhujiang Delta*. Hong Kong: The Chinese University Press, pp.271-291.

Liu, Junde and Shu, Qing (1996) "Zhongguo quyujingji de xinshijiao: Xingzhenqu jingji (A new point of view in China's regional economy: Administrative-zone economy)", *Gaige Yu Zhanlue (Reform and Strategy)*, No.5, pp.1-4.

Liu, Shenghe (1991) "Wuguo zhouqixing jishi yu xiangcun fazhan yanjiu (China's periodic market and rural development)", *Jingji Dili (Economic Geography)*, No.1, pp.79-84.

Loo, B. P. Y. (1999) "Development of a regional transport infrastructure: Some lessons from the Zhujiang Delta, Guangdong, China", *Journal of Transport Geography*, Vol.7, pp.43-63.

Lösch, A. (1954) *The Economics of Location*. New Haven: Yale University Press.

Lu, Ding (1994) *Entrepreneurship in Suppressed Markets: Private-sector Experience in China*. New York: Garland Publishing, Inc.

Ma, Chunsun (1996) "Zhongguo xinzhenquhua tizhi gaige shexiang (Some ideas on reforming China's regional administrative system)", *Zhongguo Fangyu (China's Region)*, No. 5, pp.17-21.

Mallee, H. (1995) "China's household registration system under reform", *Development and Change*, Vol.26, pp.1-29.

Marshall, J. U. (1964) "Model and reality in central place studies", *Prof. Geogr.*, Vol.16, pp.5-8.

Marton, A. M. (2000) *China's Spatial Economic Development: Restless Landscape in the Lower Yangzi Delta*. London, New York: Routledge.

Massey, D. (1985) "New directions in space", in Gregory, D. and Urry, J. (eds) *Social Relations and Spatial Structures*. London: Macmillan, pp. 9-19.

Mastel, G. (1997) *The Rise of the Chinese Economy: The Middle Kingdom Emerges*. New York: M. E. Sharpe.

McGee, T. G. (1989) "*Urbanisasi* or *kotadesasi*? Evolving patterns of urbanisation in Asia", in Costa, F. J. (eds) *Urbanisation in Asia: Spatial Dimensions and Policy Issues*. USA: University of Hawaii Press, pp.93-108.

McGee, T. G. (1991) "The emergence of desakota regions in Asia: Expanding a hypothesis", in Ginsburg, N. *et al* (eds) *The Extended Metropolis: Settlement Transition in Asia*. Honolulu: University of Hawaii Press, pp.3-25.

McGee, T. G. (1998) "Globalisation and rural-urban relations in the developing world", in Lo, Fu-chen and Yeung, Yue-man (eds) *Globalisation and the World of Large Cities*. Japan: The United Nations Universities.

McMillan, J. and Naughton, B. (1992) "How to reform a planned economy: Lessons from China", *Oxford Review of Economic Policy*, Vol.8, pp.130-143.

Mei, Maofa (1998) "Nongchanpin mainan chengyin fenxi (An analysis of the underlying causes for agricultural products' selling difficulties)", *Nongcun Jingji Yingjiu Cenkou (Reference on Rural Economy Study)*, No.2, p.25.

Min, Yaoliang (1995) "Goujian nongzhanpin shichang tixi de jige wanti (Several problems on constructing agricultural product market system)", *Zhongguo Nongcun Jingji (Chinese Rural Economy)*, No. 9, pp.10-15.

Muldavin, J. (1998) "The limits of market triumphalism in rural China", *Geoforum*, Vol.28, pp.289-312.

Murrell, P. (1992) "Evolutionary and radical approach to economic reform", *Economics of Planning*, Vol. 25, pp.79-95.

Nathan, A. J. (1997) *China's Transition*. New York: Columbia University Press.

Naughton, B. (1995a) "Cities in the Chinese economic system: Changing roles and conditions for autonomy", in Davis, D. S., Kraus, R., Naughton, B. and Perry, E. J. (eds) *Urban Spaces in Contemporary China*. New York: Cambridge University Press, pp.61-89.

Naughton, B. (1995b) *Growing Out of the Plan: Chinese Economic Reform, 1978-1993*. New York: Cambridge University Press.

Nee, V. (1992) "Organisational dynamics of market transition: Hybrid forms, property rights, and mixed economy in China", *Administrative Science Quarterly*, Vol.37, pp.1-27.

Nee, V. and Su, Sijin (1996) "Institutions, social ties and commitment in China's corporatist transformation", in Naughton, B. and McMillan, J. (eds) *Reforming Asian Socialism: The Growth of Market Institutions*. USA: The University of Michigan Press, pp.111-134.

Nolan, P. (1995) *China's Rise, Russia's Fall: Politics, Economics and Planning in the Transition from Stalinism*. New York: St. Martin's Press.

Nongmin Rebao (Peasant's Daily) (1998, June 10) *"Jinnian liangshi maigei shui (Who should peasants sell their grains to?)"*, p.1.

Nongmin Rebao (Peasant's Daily) (1998, June 26) *"Jiangning qu de baohujia baohuoli shui. (To whom is Jiangning district's planned price protecting?)"*, p.1.

O'Brien, K. J. and Li, Lianjiang (1999) "Selective policy implementation in rural China", *Comparative Politics*, Vol.31, pp.167-186.

Oi, J. C. (1992) "Fiscal reform and the economic foundations of local state corporatism", *World Politics*, Vol.45, pp.99-126.

Oi, J. C. (1995) "The role of the local state in China's transitional economy", *The China Quarterly*, Vol.144, pp.1132-1149.

Oi, J. C. (1999a) "Two decades of rural reform in China: An overview and assessment", *China Quarterly*, No. 159, pp.616-628.

Oi, J. C. (1999b) *Rural China Takes Off*. Berkeley: University of California Press.

Owen, W. (1959) "Special problems facing underdeveloped countries: Transportation and economic development", *American Economic Review*, Vol. 49, No. 2, pp.179-187.

Owen, W. (1968) *Distance and Development: Transport and Communication in India*. Washington: The Brookings Institution.

Passi, A. (1991) "Deconstructuring regions: Notes on the scales of spatial life", *Environment and Planning A*, Vol.23, pp.239-256.

Peng, Xiaoyong (1999) "Dui kaitou wuguo nongcun shichang de jianyi (Suggestions on developing China's rural markets)", *Nongye Jingji Wanti (Problems on Rural Economy)*, No. 9, pp.28-29.

Plattner, S. M. (1976) "Periodic trade in developing areas without markets", in Smith, R. T. H. (ed.) *Market-Place Trade – Periodic Markets, Hawkers and Traders in Africa, Asia, and Latin America*. Vancouver: University of British Columbia, pp.69-89.

Porter, G. (1995) "The impact of road construction on women's trade in rural Nigeria", *Journal of Transport Geography*, Vol.3, pp.3-14.

Potter, R. B. and Unwin, T. (1995) "Urban-rural interaction: Physical form and political process in the third world", *Cities*, Vol.12, pp.67-73.

PRC-MDT (People's Republic of China, Ministry of Domestic Trade) (1998) *Zhongguo Guonei Maoyi Nianjian (Almanac of China's Domestic Trade)*. Beijing: Zhongguo Guonei Maoyi Nianjian Chubanshe.

Pu, Shansin (1995) "Tui shilingdaoxian tizhi de fansi (Rethinking the system of city-leading-counties)", *Zhongguo Fangyu (China's Region)*, No.5, pp.2-7.

Pu, Xingzhu and Zhu, Qianwei (1993) *Diandai Zhongguo Xingzheng (Administration in Contemporary China)*. Shanghai: Fudan Daxue Chubanshe.

Qian, Yingyi and Stiglitz, J. (1996) "Institutional innovations and the role of local government in transition economy: The case of Guangdong province of China", in Naughton, B. and McMillan, J. (eds) *Reforming Asian Socialism: The Growth of Market Institutions*. USA: The University of Michigan Press, pp. 175-193.

Rawski, T. G. (1999) "Reforming China's economy: What have we learned", *The China Journal*, No. 41, pp.139-156.

Ren, F. and Du, Q.H. (2000) "Shijianya moshi yu zhongguo chuantongshichen yanjiu (A study of Skinner's model and China's traditional marketing towns)", *Zhejiang Shehui Kexue (Zhejiang Social Science)*, No.5, pp.111-115.

Renmin Rebao (People's Daily) (1998, October 19) *"Zhonggong zhongyang guanyu nongye he nongcun gongzuo ruogan zhongdaiweiti de jueding (Decisions of Central Committee of CCP on a considerable number of significant agricultural and rural related problems)"*, p.1.

Rowlands, M.J. (1973) "Modes of exchange and incentives for trade: With reference to later European history", in Renfrew, C. (ed.) *The Explanation of Culture Change: Models in Prehistory*. Pittsburgh: university of Pittsburgh Press, pp.589-600.

Rozelle, S. (1996) "Gradual reform and institutional development: The key to success of China's agricultural reforms", in McMillan, J. and Naughton, B. (eds) *Reforming Asian Socialism: The Growth of Market Institutions*. USA: University of Michigan Press, pp. 197-220.

Sands, B. and Myers, R. H. (1986) "The spatial approach to Chinese history: A test", *Journal of Asian Studies*, Vol.45, pp.721-743.

Sangren, P. S. (1980) *A Chinese Marketing Community: An Historical Ethnography of Ta-Chi, Taiwan*. Unpublished Ph.D. Thesis. Department of Anthropology, Stanford University. United States of America.

Sangren, P. S. (1985) "Social space and the periodisation of economic history: A case from Taiwan", *Comparative Studies in Society and History*, Vol. 27, pp.531-561.

Sayer, A. (1985) "The difference that space makes", in Gregory, D. and Urry, J. (eds) *Social Relations and Spatial Structures*. London: Macmillan, pp. 49-66.

Sayer, A. (1992) *Method in Social Science: A Realistic Approach*. London: Routledge.

Schwimmer, B. (1976) "Periodic market and urban development in Southern Ghana", in Smith, R. T. H. (ed.) *Market-Place Trade – Periodic Markets, Hawkers and Traders in Africa, Asia, and Latin America*. Vancouver: University of British Columbia, pp.123-145.

Selden, M. (1989) "Mao Zedong and the political economy of Chinese development", in Dirlik, A. and Meisner, M. (eds) *Marxism and The Chinese Experience*. New York: M. E. Sharpe, Inc.

Sethi, J. D. (1994) *Dichotomy to Continuum: Restructuring Rural-Urban Relations*. New Deli: Har-Anand Publication.

Shen, Liren (1999) *Difang Zhenfu De Jingji Zhineng He Jingji Xingwei (Local Government's Economic Functions and Behaviour)*. Shanghai: Yuandong Chubanshe.

Sheng, Taiji and Chen, Zhongwei (1996) "Lun wuguo nongcun shichang peiyuzhong de zhenfu xingwei he tiaokong jizhi (Discussions on governmental behaviours and regulation mechanisms during the growing of rural market system)", *Zhongguo Nongcun Jingji (Chinese Rural Economy)*, No.9, pp.43-47.

Sheppard, E. (1995) "Dissenting from spatial analysis", *Urban Geography*, Vol.16, pp.283-303.

Shi, Yishao (1995) *Zhongguo Nongcunjishi De Lilun yu Shiqian (Theory and Practice of China's Rural Markets)*. Shanxi: Renmin Chubanche.

Shu, Weiguo (1999) "Kaituo nongcunshichang shi kuodai guoneixuqiu ladong jingjizengzhang de yige zhongyao jucuo (Expanding rural markets is a crucial measure to boost domestic demand and achieve economic growth)", *Zhongguo Nongcun Jingji (Chinese Rural Economy)*, No.3, pp.36-40.

Shue, V. (1988) *The Reach of the State: Sketches of the Chinese Body Politic*. Stanford: Stanford University Press.

Shue, V. (1995) "State sprawl: the regulatory state and social life in a small Chinese city", in Davis, D. S., Kraus, R., Naughton, B. and Perry, E. J. (eds) *Urban Spaces in Contemporary China*. New York: Cambridge University Press, pp.90-112.

Skinner, G. W. (1964) "Marketing and social structure in rural China, Part I", *Journal of Asian Studies*, Vol.24, pp.3-43.

Skinner, G. W. (1965a) "Marketing and social structure in rural China, Part II", *Journal of Asian Studies*, Vol.24, pp.195-228.

Skinner, G. W. (1965b) "Marketing and social structure in rural China, Part III", *Journal of Asian Studies*, Vol.24, pp.363-399.

Skinner, G. W. (1977a) "Urban development in Imperial China", in Skinner, G. W. (ed.) *The City in Late Imperial China*. Stanford: Stanford University Press, pp.3-31.

Skinner, G. W. (1977b) "Regional urbanisation in nineteenth-century China", in Skinner, G. W. (ed.) *The City in Late Imperial China*. Stanford: Stanford University Press, pp.211-249.

Skinner, G. W. (1977c) "City and the hierarchy of local systems", in Skinner, G. W. (ed.) *The City in Late Imperial China*. Stanford: Stanford University Press, pp.275-351.

Skinner, G. W. (1985a) "Rural marketing in China: Revival and reappraisal", in Plattner, S. (ed.) *Markets and Marketing*. Lanham: University Press of America, pp.7-47.

Skinner, G. W. (1985b) "Rural marketing in China: Repression and revival", *The China Quarterly*, No.103, pp.393-413.

Skinner, G. W. (1994) "Different development in Lingnan", in Lyons, T. P. and Nee, V. (eds) *The Economic Transformation of South China: Reform and Development in the Post-Mao Era*. New York: East Asia Program, Cornell University, pp.17-54.

Smith, C. A. (1974) "Economics of marketing system: Models from economic geography", *Annual Review of Anthropology*, Vol. 3, pp.167-201.

Smith, C. A. (ed.) (1976a) *Regional Analysis. Volume One: Economic Systems*. London: Academic Press.

Smith, C. A. (ed.) (1976b) *Regional Analysis. Volume Two: Social Systems*. London: Academic Press.

Smith, R. H. T. (1979) "Periodic market-places and periodic marketing: Review and prospect – I", *Progress in Human Geography*, Vol.3, pp. 472-505.

Smith, S. J. (1999) "Society-space", in Cloke, P. Crang, P. and Goodwin, M. (eds) *Introducing Human Geographies*. London and New York: Arnold, pp.12-23.

Soja, E. (1996) *Thirdspace*. Oxford: Blackwell.

Spencer, J. E. (1940) "The Szechuen village fair", *Economic Geography*, Vol.16, pp.48-58.

SSB (State Statistical Bureau) (1989) *Fenjin De Sishinian (Exertion For Forty Year1949-1989)*. Beijing: Zhongguo Tongji Chubanshe.

SSB (State Statistical Bureau) (1998) *Zhongguo Nongcun Tongji Nianjian 1998 (Rural Statistical Yearbook, 1998)*. Beijing: Zhongguo Tongji Chubanshe.

SSB (State Statistical Bureau) (1999) *Zhongguo Nongcun Tongji Nianjian 1999 (Rural Statistical Yearbook, 1999)*. Beijing: Zhongguo Tongji Chubanshe.

SSB (State Statistical Bureau) (2001) *Zhongguo Tongji Nianjian 2001 (Chinese Statistical Yearbook 2001)*. Beijing: Zhongguo Tongji Chubanshe.

SSB-GMSD (State Statistical Bureau, Goods and Materials Statistic Department) (1999) *Zhongguo Shichang Tongji Nianjian (China's Market Statistic Yearbook, 1998)*. Beijing: Zhongguo Tongji Chubanshe.

SSB-RSST (State Statistic Bureau – Rural Socio-economic Survey Team) (2002) *Zhongguo Nongcun Tongji Nianjian 2002 (Chinese Rural Statistical Yearbook 2002)*. Beijing: Zhongguo Tongji Chubanshe.

SSB-Well-off Study Group (1992) *Zhongguo Xiaokang Biaozhun (Criteria on China's Well-Off Status)*. Beijing: Zhongguo Tongji Chubanshe.

Sunley, P. (1996) "Context in economic geography: the relevance of pragmatism", *Progress in Human Geography*, Vol.20, pp.338-355.

Symanski, R. (1978) "Periodic markets in Southern Colombia", in Smith, R. T. H. (ed.) *Market-Place Trade – Periodic Markets, Hawkers and Traders in Africa, Asia, and Latin America*. Vancouver: University of British Columbia, pp. 171-185.

Szymanski, R. and Agnew, J.A. (1981) *Order and Skepticism: Human Geography and the Dialectic of Science*. Washington: Association of American Geographers.

Tang, W.-S. (1995) *Urbanisation in China's Fujian Province Since 1978*. Occasional Paper No.43. Hong Kong Institute of Asia-Pacific Studies, The Chinese University of Hong Kong.

Tang, W.-S. (1997) "Urbanisation in China: A review of its causal mechanisms and spatial relations", *Progress in Planning*, Vol.48, pp.1-65.

Tang, W.-S. and Chung, H. (2000) "Urban-rural transition in China: Beyond the desakota model", in Li, Si-Ming and Tang, W.-S. (eds) *China's Regions, Polity and Economy: A Study of Spatial Transformation in the Post-Reform Era*. Hong Kong: The Chinese University of Hong Kong Press, pp.275-308.

Tang, W.-S. and Chung, H. (2002) "Rural-urban transition in China: Illegal land use and construction", *Asia Pacific Viewpoint*, Vol.43, pp.43-62.

Tang, W.-S. and Lee, W.Y. (2003) *Some Reflections of Modern Geographic Thoughts on China Urban and Regional Studies*. Hong Kong: Occasional Paper 25, The Center for China Urban and Regional Studies, Hong Kong Baptist University.

Unger, J. and Chan, A. (1996) "Corporatism in China: A Developmental State in an East Asian Context", in McCormick, B. L. and Unger. J. (eds) *China After Socialism: in the Footsteps of Eastern Europe or East Asia*. New York: M.E. Sharpe, pp.95-129.

Unwin, T. (1989) "Urban-rural interaction in development countries: A theoretical perspective", in Potter, R. B. and Unwin, T. (eds) *The Geography of Urban-Rural Interaction in Developing Countries*. London: Routledge, pp.11-32.

Urry, J. (1985) "Social relations, space and time", in Gregory, D. and Urry, J. (eds) *Social Relations and Spatial Structures*. London: Macmillan, pp.20-48.

URSED-Study Group (Chengxiang Eryuanjiegouxia Jingjishehui Xietiaofazhan Ketizu) (1996a) "Zhongguo chengxiang jingji ji shihui de xietiaofazhan (China's urban-rural socio-economic development, Part I)", *Guanli Shijie (Management World)*, No. 3, pp.15-24.

URSED-Study Group (Chengxiang Eryuanjiegouxia Jingjishehui Xietiaofazhan Ketizu) (1996b) "Zhongguo chengxiang jingji ji shihui de xietiaofazhan (China's urban-rural socio-economic development, Part II)", *Guanli Shijie (Management World)*, No. 4, pp.35-44.

Walder, A. (1995) "Local government as industrial firms: An organisational analysis of China's transitional economy", *American Journal of Sociology*, Vol.101, pp.263-301.

Wang, Guihuan (1994) *Zhongguo Nongcun Xindaihua Yu Nongmin (China's Rural Modernization and Peasants)*. Guizhou: Renmin Chubanshe.

Wang, M. Y. L. (1997) "The disappearing rural-urban boundary: Rural socioeconomic transformation in the Shenyang-Dalian region of China", *Third World Planning Review*. Vol.19, pp.229-250.

Wang, Mingming (1997) *Shihui Renleixue Yu Zhongguo Yingjiu (Social Anthropology and It's Studies of China)*. Beijing: Sanlian Shudian.

Wang, Shaoguang and Hu, Angang (1999) *The Political Economy of Uneven Development: The Case of China*. New York: M. E. Sharpe, Inc.

Wang, Weiqi (1992) "Zhonggong defang zhengquan shiguanxian tizhi de xingcheng yu fazhan (Administrative changes above county-level in mainland China since 1978)", *Zhongguo Dalu Yanjiu (Mainland China Studies)*, Vol.35, pp.22-38.

Watson, A. (1988) "The reform of agricultural marketing in China since 1978", *China Quarterly*, No.113, pp.1-28.

Watson, A. (1992) "The management of the rural economy: The institutional parameters", in Watson, A. (ed.) *Economic Reform and Social Change in China*. New York: Routledge, pp.171-199.

Watson, A. (1996) "Conflict over cabbages: the reform of wholesale marketing", in Garnaut, R., Guo, Shutain and Ma, Guonan (eds) *The Third Revolution in the Chinese Countryside*. Cambridge: Cambridge University Press, pp.144-163.

Watson, A. and Findlay, C. (1992) "The 'wool war' in China", in Findlay, C. (ed.) *Challenges of Economic Reform and Industrial Growth: China's Wool War*. Australia: Allen & Unwin, pp.163-180.

Watson, A., Findlay, C. and Du, Yintang (1989) "Who won the 'wool war'?: A case study of rural product marketing in China", *China Quarterly*, No.118, pp.213-241.

Wedeman, A. (1997) "Stealing from the farmers: Institutional corruption and the 1992 IOU crisis", *China Quarterly*, No.152, pp.805-831.

Wood, L. J. (1978) "Rural markets in Kenya", in Smith, R. T. H. (ed.) *Market-Place Trade – Periodic Markets, Hawkers and Traders in Africa, Asia, and Latin America*. Vancouver: University of British Columbia, pp. 222-239.

World Bank (1985) *China: The Transport Sector*. Washington: The World Bank.

World Bank (1990) *China: Between Plan and Market*. Washington: The World Bank.

World Bank (1992) *China: Reform and the Role of the Plan in the 1990s*. Washington: The World Bank.

World Bank (1994) *China: Internal Market Development and Regulation*. Washington: The World Bank.

Wu, Chengming (1983) "Lun Mingdai guonei shichang he shengren zhiban (Discussion on the internal market and entrepreneurial capital during the Ming dynasty)", in Zhongguo Shihui Kexueyuan Jingji Yanjiusuo Xuexu Weiyuanhui (ed.) *Zhongguo Shihui Kexueyuan Jingji Yanjiusuo Jikan, Di wu juan (Journal of Economic Studies Centre, Chinese Academy of Social Science)*, Vol. 5, pp.1-32.

Xia, Rushan and Yang, Huantang (1991) "Guanyu shixing shiguanxian tizhi de tantao (A discussion in the adoption of city-leading-counties)", in The Chinese Administrative Planning Study Committee (ed.) *Zhongguo Xinzheng Quhua Yanjiu (Studies of Chinese Administrative Planning)*. Beijing: Zhongguo Shehui Chubanshe.

Xiao, Yibin (1998) "Maifang shichang xia nongmin xuqiu de shizheng fenxi (A positive study of farmers' demands under circumstances of a buyer's market)", *Zhongguo Nongcun Guanci (China Rural Survey)*, No.6, pp.46-53, 62.

Xie, Dixiang (1998) "Nongcun jimaoshichang jianshe: yiyi, wenti ji fazhanduici (Establishing rural markets: meaning, problems and strategies)", *Zhongguo Liutong Jingji (China Business and Market)*, No.4, pp.4-7.

Xu, Boyuan (1996) "Agricultural wholesale markets", in Garnaut, R. and Ma, Guonan (eds) *The Third Revolution in the Chinese Countryside*. United Kingdom: Cambridge University Press, pp.113-119.

Yabuki, Susumu (1999) *China's New Political Economy*. Translated by Harner, S. M. USA: Westview Press.

Yan, Xianbo (2002) "Zongxin shenshi zhongguo nongcunshicheng yu nongmin xiaofei (Re-examining Chinese rural markets and peasant consumption", *Jingji Yanjiu Cankou (Reference on Economic Studies)*, No.25, pp.25-28.

Yang, Chian-kun (1944) *A North China Local Market Economy: A Summary of a Study of Periodical Markets in Chowping Hsien, Shantung*. New York: Institute of Pacific Relations.

Yang, Chun (1996) "The changing urbanisation of Pearl River Delta: From a planned economy to a world market economy", in Li, S. M., Tang, W.-S., Jiang, Lanhong and Zhou, Suqing (eds) *Perspectives on China's Regional Economy*. Taipei: Population Institute, pp.77-103. (In Chinese).

Yang, D. L. (1997) *Beyond Beijing: Liberalisation and the Regions in China*. New York: Routledge.

Yeh, A. G. O. (2000) "Foreign investment and urban development in China", in Li, Si-Ming and Tang, W.-S. (eds) *China's Regions, Polity and Economy: A Study of Spatial Transformation in the Post-Reform Era*. Hong Kong: The Chinese University of Hong Kong Press, pp.35-64.

Zhang, Xiaohe, Lu, Weiguo, Sun, Keliang, Findlay, C. and Watson, A. (1991) *The "Wool War" and the "Cotton Chaos": Fibre Marketing in China*. Working Paper No. 91/14. Chinese Economy Research Unit. The University of Adelaide.

Zheng, Fuhua (1999) "Butong shouru cengci nonghu xiaofei chaiyi fenxi (An analysis of consumption variation between peasants of different income level)", *Nongye Jingji Wenti (Problems on Rural Economy)*, No.10, pp.24-27.

Zheng, Xinmin and Zheng, Xiaoshan (1998) "Tigao goumaili, kaitou nongcun shichang (Increasing purchasing power so as to develop rural market)", *Zhongguo Nongcun Jingji (Chinese Rural Economy)*, No.4, pp.4-6.

Zheng, Xueying (1997) "1997-1998 nian xiaofeiping shichang fenxi yu yuce (An analysis and projection of the consumer goods market, 1997-1998)", in Guojia Xinxizhongxin (eds) *Zhongguo Shichang Zhenwang (Prospects on China's Markets)*. Beijing: Zhongguo Jihua Chubanshe, pp.15-21.

Zhou, H. (1994) "Zongliang pingheng shichang gouxiao (Keeping the aggregate amount in balance and practising market purchase)", *Nongye Jingji Wenti (Problems of Agricultural Economy)*, No.5, pp.10-15.

Zhou, Yixing (1991) "The metropolitan interlocking region in China: A preliminary hypothesis", in Ginsburg, N., Kopper, B. and McGee, T. G. (eds) *The Extended Metropolis: Settlement Transition in Asia*. Honolulu: University of Hawaii Press, pp.89-111.

Zhou, Zhendong (1998) "Guanyu nongmin zengshou wenti de sikou (Considering problems in increasing peasants' income)", *Nongcun Jingji Yingjiu Cenkou (Reference on Rural Economy Study)*, No.3, pp.15-17.

Zhu, Yuxiang (1997) *Zhongguo Jingdai Nongmin Wanti Yu Nongcun Shihui (Problems of Modern China's Peasant and Rural Society)*. Shangdong: Shangdong Renmin Chubanshe.

ZQSB (Zhaoqing Statistical Bureau) (1997) *Zhaoqing Tongji Nianjian 1997 (Zhaoqing Statistical Yearbook 1997)*. Unpublished material.

ZQSB (Zhaoqing Statistical Bureau) (1998) *Zhaoqing Tongji Nianjian 1998 (Zhaoqing Statistical Yearbook 1998)*. Unpublished material.

ZQSB (Zhaoqing Statistical Bureau) (2000) *Zhaoqing Tongji Nianjian 2000 (Zhaoqing Statistical Yearbook 2000)*. Unpublished material.

ZQSB (Zhaoqing Statistical Bureau) (2002) *Zhaoqing Tongji Nianjian 2002 (Zhaoqing Statistical Yearbook 2002)*. Unpublished material.

Zweig, D. (1992) "Urbanisation Rural China: Bureaucratic Authority and Local Autonomy", in Lieberthal, K. G. and Lampton, D. M. (eds) *Bureaucracy, Politics and Decision Making in Post-Mao China*. California: University of California Press, pp.334-363.

Zweig, D. (1997) *Freeing China's Farmers: Rural Restructuring in the Reform Era*. New York: M. E. Sharpe.

Index

administrative hierarchy 15, 29, 33, 37, 41-42, 57-58, 122, 150-151, 159
administrative parameters 12, 21, 89, 121
administrative-zone economy 11, 41, 44, 48-49, 122, 132
artificial system 25, 42

birdcage economy 2
Blecher, M. and Shue, V. 33, 39

Cao, Tiandian 45, 56, 165
causal power 7-8
cellular structure 41, 147
central place theory
 as mountain model 25-26, 60, 63, 161
 as plains model 25-26, 60, 63, 161
 by Skinner's modification 21-26
Chinese Communist Party 9, 56
Christaller, W.
 on central place theory 22-24
Chung, H.
 on Chinese governance 37-38, 146
 on decentralisation 49
 on toll stations 140
Chung, H. and Tang, W-S.
 on cities organisation 21
 on decentralisation 5, 33
city hierarchy
 central metropolis 28
 greater city 28
 intermediate market town 25
 local city 28
 regional city 28
 regional metropolis 28
 standard market town 25
city-leading-counties 38, 151, 157
commodity war 6, 32, 41, 58, 132
contextual approach 6-7
core-periphery structure 151-159
Crissman, L.W. 5, 70

decentralisation 5-6, 11, 19, 37-40, 45, 49, 153

Deqing
 administrative control 151-154
 basic characteristics 13-18, 61-69
 income level and shopping frequency 102-108
 market days 74
 market hierarchy 70-84, 89-90
 purchase price and trade barriers 132-138
 self-sufficiency 91-102
 single market structure 122-131
 toll station 140, 144-146
 transport 108-120
 urban status 159-160
desakota 154
Dongguan
 administrative control 151-154
 administrative principle on markets 126-131
 basic characteristics 13-18, 61, 69
 income level and shopping frequency 102-108
 market days 77, 84
 market hierarchy 70-81, 84-90
 purchase price and trade barriers 132-138
 self-sufficiency 91-102
 toll station 144, 146
 transport 108-110, 113-120
 urban status 157-158
double-track system 31, 38, 165
Duncan, S. 7-8

economic reform 2-5, 31-34, 35, 132-133, 157-158, 163-164

Findlay, C. Watson, A. and Martin, W. 3, 4, 6, 31-33
flooding 99-101

government regulations 132
 impact 136
 production of speciality items 138
 purchase price 133, 136-138
 trade barriers 132, 136, 138
Great Leap Forward 29, 70

half-baked reform 41, 45
Han, Jin 58, 102
Hodder, R. 5, 72, 90, 119
household registration system 37, 42, 146, 157

Imperial China 3, 25, 28, 42
income level and shopping frequency 102-108
Industry and Commerce Management Bureau (ICMB) 38, 50-51, 68, 77, 78, 84, 85, 125
intensification cycle 26-27
integration 21, 29, 41, 44, 56, 128, 155
invisible hand 122, 127, 139, 147, 165-166
invisible wall 21, 138, 139-141, 147, 164-166

Koppel, B. 154-156
Koppel, B. and Hawkins, J. 154-156

land requisition 49, 50, 96-102, 127, 137
Lin, G.C.S. 21, 108-109
Liu, Junde and Shu, Qing 21, 45
local
- causal process 8
- corporatism 39, 41, 158
- particularism 30

locality effect
- in rural China 91-120
- in theory 8

lunar calendar 25, 51

Mao's era 8, 21, 35
market
- market co-operatives 127
- market cycles 25, 28
- market fees 1, 125-126

market hierarchy
- central market 24, 28-29
- central market town 25
- intermediate market 25, 28-29
- periodic market 2-3
- single market structure 122, 125, 127
- standard market 24, 28-29
- wholesale market 6, 10, 29, 37, 50-59, 64, 66, 81, 88-89, 164

Marton, A.M. 21, 33, 40
methodical deradicalisation 31, 121
Muldavin, J. 37, 164, 166

Nathan, A.J. 8-9

Oi, J.C. 33, 39, 165

Passi, A. 153, 156
People's Commune 29-33, 35, 36, 39, 42
planned prices 40
political-cum-economic hierarchy 42-45, 49-50, 56-59, 89, 146-147, 156
privatisation 9, 38
procurement quota 57-58, 66, 132

Rawski, T.G. 9, 36
region
- institutionalisation of region 153-154
- regional identity 153
- regional inequality 150

revenue 38-41, 44, 48, 63, 69, 132, 153, 155
Roadway Bureau 68, 99, 139

Sayer, A. 7-8
Selden, M. 8-9
self-sufficiency 9, 19, 63-72, 91-102, 161-162
Shi, Yishao 3-4, 50, 57
shortage economy 152
Shu, Weiguo 4, 41
Shue, V. 33, 37, 39
Skinner, G.W.
- modelling with central place theory 21-26, 60, 63, 163
- on Chinese market system 2, 5, 21-34, 58, 67
- on core-periphery concept 150-153, 159
- on market hierarchy 25, 28, 41, 48, 50, 72, 81, 88
- on persistent cycle of policy 28, 121
- on self-sufficiency 27, 91-102
- on transport, 108-109 119-120, 143

socialism with Chinese characteristics
- in agricultural sector 8-11

space 7-8, 153-160

absolute space 7
locality effect 8
spatial boundary effect 8
Spatial contingency effect 8, 35-45
Third space 159-160
specialisation 10, 16, 61-90, 96, 101, 118, 124-130, 161
subcontractor 132-133, 138

Tang, W-S.
on Chinese governance 37-38
on decentralisation of power 5
on shortage economy in China 152
on space 7-8
Tang, W-S. and Chung, H.
on Chinese governance 147
on decentralisation of power 38
on economic integration 21, 156
transport
bicycles 110
market vessels 60, 74, 84, 109,117
toll stations 19, 69, 139-146, 152

urban-rural dichotomy 154-156

Waston, A. 3, 4, 33
white slips 17, 49
World Bank 31, 32, 37, 67
World Trade Organisation 2, 57

Xu, Boyuan 54, 56

Zhaoqing 62, 66, 68
Zweig, D. 33, 37